Praise for

Essential Natural Plasters

More than a book on plasters, *Essential Natural Plasters* delves deep into the world of natural building, explaining the detailing and function of walls finished with natural plasters in addition to the fundamentals of plasters made from earth, lime, and gypsum. It covers the spectrum of natural plasters, providing fundamental advice on mixing, application, and finishing, but also discusses often overlooked details such as masking and maintenance. Including tons of recipes, *Essential Natural Plasters* should be in the hands of every natural builder.

— Kyle Holzhueter, PhD, First Degree Certified Plasterer, Permaculture Center, Kamimomi, Japan

This is hands down the most comprehensive book on natural plasters that I've ever read. The authors include information you'd expect in a textbook, yet they write in a clear, easy-to-read style. Everything you'd want to know about natural plaster is included — from preparation and planning, through choosing materials, application techniques, and even recipes. *Essential Natural Plasters* is indeed essential for every natural builder's library!

— Sigi Koko, Down to Earth Designs, www.buildnaturally.com

From an engineer's point-of-view, plaster can be both a weak point and a strength in a natural material wall assembly. Tina and Michael, supported by their myriad plastering guru contributors, have curated a wonderfully broad and incredibly detailed recipe book for anyone who wants to work with natural plasters. There is background on a wide variety of materials, wisdom from many projects, and inspiration galore here. For anyone from beginner wall finisher to expert plasterer, this is a valuable resource — one that moves the state of the art forward here in North America.

— Tim Krahn, P. Eng., structural engineer at Building Alternatives Inc.

Essential Natural Plasters: A Guide to Materials, Recipes, and Use is aptly named. Tina Therrien and Michael Henry have delivered a thorough and honest easy-to-read guide for anyone interested or working in the natural plaster field. This book guides the reader through the stages of plaster work, beginning with onsite and personal safety and closing with much-needed coverage calculations that only a professional would know. They have given the reader an unprecedented 27 plaster recipes from around the world developed by professionals to help guide others in the complex and temperamental field of working with natural materials. The information in the book will save others years of personal study and experimentation. I wish I had this book 20 years ago when I started out!

— Janine Bjornson, Natural Builder, Educator, and Consultant

Tina Therrien and Michael Henry have created a superb, comprehensive, and well-illustrated guide to natural plasters. Drawing on their own extensive experience, and the experience and wisdom of other leaders in this field, they have woven together a treasure trove of practical and insightful information about materials, processes, decision-making, recipes, tricks of the trade, and essential practices.

— David Eisenberg, Director, Development Center for Appropriate Technology

Essential Natural Plasters is the book I've been waiting for — clear, concise, and thorough knowledge delivered in a thoughtfully organized format. Finally, a book for plasterers by plasterers with a fountain of knowledge on the subject and a killer selection of recipes, to boot. This is easily my new go-to resource for all things natural plaster in North America.

— Ziggy Liloia, owner, instructor, builder, The Year of Mud, www.theyearofmud.com

Essential Natural Plasters is definitely an essential book for anyone wanting to understand the complex art of mixing and applying appropriate natural plasters both inside and outside their building project. The authors have drawn on their extensive experience as professional plasterers, as well as the expertise of numerous others well-versed in the art of plastering. The myriad recipes for specific plasters for virtually every application is well worth the price of the book alone.

— Kelly Hart, author, *Essential Earthbag Construction* and founder, www.greenhomebuilding.com

Born-again natural builders, in these pages you can hear the voices of the past researcher and former educator who are the authors. *Essential Natural Plasters* is one part carefully detailed cookbook to three parts practical, shared experience, blended deftly with an admixture of inspiration and a dash of humor. Together, Tina and Michael reveal the great secret recipe that is natural plastering: experiment, fail, repeat, enjoy!

— Ben Polley, co-founder, Evolve Builders Group Inc, founder, Fermata — Works of Earth

essential NATURAL PLASTERS

a guide to materials, recipes, and use

Michael Henry & Tina Therrien

Series editors
Chris Magwood and Jen Feigin

Title list
Essential Hempcrete Construction, Chris Magwood
Essential Prefab Straw Bale Construction, Chris Magwood
Essential Building Science, Jacob Deva Racusin
Essential Light Straw Clay Construction, Lydia Doleman
Essential Sustainable Home Design, Chris Magwood
Essential Cordwood Building, Rob Roy
Essential Earthbag Construction, Kelly Hart
Essential Natural Plasters, Michael Henry & Tina Therrien

See www.newsociety.com/SBES for a complete list of new and forthcoming series titles.

THE SUSTAINABLE BUILDING ESSENTIALS SERIES covers the full range of natural and green building techniques with a focus on sustainable materials and methods and code compliance. Firmly rooted in sound building science and drawing on decades of experience, these large-format, highly illustrated manuals deliver comprehensive, practical guidance from leading experts using a well-organized step-by-step approach. Whether your interest is foundations, walls, insulation, mechanical systems, or final finishes, these unique books present the essential information on each topic including:

- Material specifications, testing, and building code references
- Plan drawings for all common applications
- Tool lists and complete installation instructions
- Finishing, maintenance, and renovation techniques
- Budgeting and labor estimates
- Additional resources

Written by the world's leading sustainable builders, designers, and engineers, these succinct, user-friendly handbooks are indispensable tools for any project where accurate and reliable information is key to success. GET THE ESSENTIALS!

Cover design by Diane McIntosh. Cover images property of the authors.

Illustrations by Dale Brownson.
Background photo author supplied.

Printed in Canada. First printing April 2018.

Any other inquiries can be directed by mail to:
New Society Publishers
P.O. Box 189, Gabriola Island, BC V0R 1X0, Canada
(250) 247-9737

LIBRARY AND ARCHIVES CANADA CATALOGUING IN PUBLICATION

Henry, Michael, author
Essential natural plasters : a guide to materials, recipes and use / Michael Henry & Tina Therrien.

(Sustainable building essentials)
Includes index.
Issued in print and electronic formats.
ISBN 978-0-86571-870-8 (softcover).--ISBN 978-1-55092-663-7 (PDF).--
ISBN 978-1-77142-258-1 (EPUB)

1. Plaster--Handbooks, manuals, etc. 2. Plastering--Handbooks, manuals, etc. 3. Sustainable construction--Handbooks, manuals, etc. 4. Building materials--Environmental aspects. I. Therrien, Tina, author II. Title. III. Series: Sustainable building essentials

TH8135.H46 2018 693'.6 C2018-900780-X
C2018-900781-8

Funded by the Government of Canada | Financé par le gouvernement du Canada | Canada

New Society Publishers' mission is to publish books that contribute in fundamental ways to building an ecologically sustainable and just society, and to do so with the least possible impact on the environment, in a manner that models this vision.

Contents

Chapter 1

Introduction

NATURAL PLASTERS are beautiful, nontoxic to live with (though not always to work with), and steeped in tradition. The act of plastering is generally enjoyable, even addictive for some, but it's very hard work. It's also serious business — along with roof and flashing details, the plaster skin of a building protects the materials inside from degradation by water, wind, sun, and animals. The job of the natural plasterer today is to take millennia-old techniques and materials, combine them with contemporary materials and tools, and employ them safely and efficiently on a modern construction site. This book gives detailed direction on how to do this. Many natural builders have collaborated to share their expertise for this book; the variety of recipes — and the diverse approaches to plastering they reflect — make this book a valuable resource for beginner and professional alike.

Fig. 1.1: *Natural plasters are a nontoxic and beautiful finish with many benefits to the homeowner, and the planet.* CREDIT: DEIRDRE MCGAHERN

Why Use Natural Plasters

Before we launch into nine chapters on how to use natural plasters, it's worth taking a moment to reflect on why you would want to use them. There are several situations in which you'd be likely to use natural plasters: to cover a natural wall system, in which case the permeability and flexibility is important and often essential; to cover more conventional wall systems, such as a stud wall sheathed with wood lath or drywall, where natural plasters add beauty and are a nontoxic alternative to paint or other wall finishes; and in restoration, to match or repair heritage plasters.

Here's a short list of some of the advantages of natural plasters:

- In our increasingly sealed homes, indoor air quality is important, and there's a growing body of evidence that the chemicals we surround ourselves with can cause harm in relatively low concentrations. Natural plasters are free of these environmental toxins.
- Natural plasters connect us to our heritage. They have a track record going back thousands of years. We know that they work, and we know how they interact with other natural building materials, including wood. Some of this knowledge has nearly been lost, but as a natural plasterer, you can help keep this knowledge alive.
- Natural plasters have greater flexibility and vapor permeability than most synthetic materials. They tend to protect the materials they are bonded to from moisture damage. They are essential as a coating for many forms of natural building, and can be beneficial for many forms of conventional construction.
- Natural plasters typically have a low embodied energy — the energy it takes to mine, process, and transport them. They can often be sourced locally and thus contribute to the local economy.

- They are beautiful. There is evidence that human happiness is tied to our connections to the natural world, and natural plasters can contribute to human well-being by introducing natural products, forms, and textures into homes.
- Natural plasters can help regulate temperature and humidity in homes, improving comfort and reducing the need for air conditioning and heating.

How to Use This Book

While at heart this is a recipe book, to be a successful plasterer you will need to understand the materials and how they interact with environment, substrate, and design.

The opening chapters of this book describe the materials, how to design for them, how to prepare the walls, and how to mix and apply natural plasters in general. It's tempting to jump straight into putting mud on the walls, but the preparatory steps leading up to that moment are more likely to determine success or failure than the days spent plastering. Chapter 3, Planning and Preparation, is probably the most important chapter in this book.

Fig. 1.2: *Damage to the base of this wall was caused by a poor choice of exterior plaster combined with a poor design for roof drainage.* CREDIT: MICHAEL HENRY

Before you begin, you will want to make sure you have chosen the best plaster for your application and that your house is designed appropriately. Too often, we have been called in to repair plasters that weren't appropriate for the site or design of the building. This may cause the plaster to fail quickly, or — even worse — it can cause damage to the underlying building materials.

When you're ready to plaster, Chapters 5 through 8 will tell you how to process and use earth, lime, gypsum, and cement plasters, giving you recipes for a wide variety of plasters. A plaster recipe is only a starting point. When you use a recipe from this book, there will be a learning stage as you come to understand the properties of the plaster: how to lay it on the wall, how thick it can be applied, how long it needs to set up before a finishing pass, whether it needs burnishing or compression, and how many coats are needed. Much of this information can be gleaned from the recipe, but some things you'll have to learn by doing. This all becomes more complicated in the real world, where there are multiple variations: change the substrate, or use a different aggregate, or if the weather changes while you're working, and you will get different results.

Testing

Always do tests.

Make the patches large (3–4 square feet) and as thick, or thicker, than you intend to apply your plaster. Also, if possible, try it on a wall at home and live with it for a while before plastering a whole room or a whole house. Get to know the plaster and understand how it works with your locally available materials. Take detailed notes, and monitor the coverage rate. Rates in recipes, when they are given, are only guidelines.

Good notes are essential when you start modifying recipes — which will happen sooner than you expect. When you change a recipe, try to change only one thing at a time. If the plaster is cracking, try adding aggregate, or changing the type of aggregate, or add fiber, or simply apply it in a thinner coat. But only do one of these things before trying something else.

Finally, be cautious. Experiment on a wall of your own house or (preferably) an outbuilding. Use mistakes as learning opportunities. Take it seriously, but have fun too.

A Note on Measurement Units

We've tried to give imperial and metric units without making things unwieldy. In some cases, we assume a liter is the same as a quart instead of being 1.06 quarts. When in doubt, always use the *ratio* in the recipe as your starting point.

Appendix 1 has useful conversion tables.

Safety: The First Priority

Toxicity and Material Safety Data Sheets

People who are new to natural plasters sometimes think anything natural must be nontoxic: earth plasters are made out of materials dug from the ground, so how could they be dangerous? In fact, although the end result is nontoxic, these products can still be hazardous to work with. Take clay for instance, which often contains large amounts of crystalline, or "free," silica (fine quartz). When inhaled, this can cause *silicosis* (a debilitating lung disease) or lung cancer. Silica is also found in cement and fine sand, but not in pure lime (which nevertheless isn't great to breathe in). The long and the short of it is that plasterers work with materials in fine powder form and need to be very careful about what they breathe in.

- Always wear an appropriate respirator when mixing, or anytime there is dust — *including cleanup!* Don't make dust when unprotected co-workers are present.
- Use a mop or a vacuum with a HEPA filter instead of sweeping when fine plaster dust is present. Wear a mask even while vacuuming.
- Read the Material Safety Data Sheet (MSDS) for materials you are working with, including bagged clays and pigments.

An MSDS (or SDS) sheet can be found for most materials by doing an internet search. For example, searching for "msds epk" finds that EPK (Edgar Plastic Kaolin) bagged clay contains 0.1–4% crystalline silica, whereas Bell Dark ball clay contains 10–30% silica.

If you are an employer, it is your responsibility to have current MSDS sheets available on site. You must make sure everyone is adequately trained to use all material and equipment safely, and that everyone knows what to do in case of an emergency.

Fig. 1.3: *One of the most overlooked hazards on the jobsite is the dust raised by sweeping. Always wear a respirator during cleanup. A vacuum with a good filter is better than a broom.* Credit: Solomon Johnston

Pigments vary greatly in their toxicity. The composition and toxicity of pigments is discussed in Chapter 2, Natural Plaster Ingredients.

Some natural materials may not have a safety data sheet, but be careful of *any* material that can produce dust. For example, straw contains enough silica (3–5% — or more) that dust from processing straw should never be inhaled.

Choosing respirators and vacuum filters

Plasterers generally use a half-face mask respirator with rubber or silicone seals that accept replaceable filters. Filters have a NIOSH (National Institute for Occupational Safety and Health) rating that ranges from N95 to P100. *N* stands for *Not oil resistant, R* indicates *oil Resistant,* and *P* is *oil Proof.* The number 95 or 100 represents the percentage of 0.3 micron particles that are blocked by the mask in tests — so 100 is the best you can get. Oil resistance is not usually a consideration for plasterers, but filters rated as P100 (oil proof, 100% efficiency at 0.3 microns) are widely available, and they offer maximum protection against fine particles, so this is what we usually use. HEPA vacuum filters also provide close to 100% protection against fine particles, and they should be used for clean-up in combination with a respirator.

Organic respirator cartridges can be used against dust only if they have the appropriate NIOSH rating, but they are primarily designed to protect against chemicals and high-VOC products, including natural solvents such as citrus solvent or turpentine.

Site Safety

Site safety should become part of your workplace culture, and it can't be over-emphasized. Take safety seriously and invite input from everybody on site about how to make the workplace safer. Not only will this uncover problems that supervisors or team leaders may have overlooked, but group participation will raise individual awareness. Make sure workers have the necessary training. Set realistic rules, then stick to them. A single accident can have significant financial, legal, and personal implications. If you are an employer, depending on where you are located, you may be obliged to have a safety representative who checks daily to make sure that any potential job hazards are addressed and discusses any dangerous work habits with the crew.

Don't use combustion engines inside of a building, and if you are doing cold-weather plastering, make sure to have proper ventilation if propane heaters are used. Plan for proper ventilation when spraying *any* material or when dust is created. Ensure that all crew members have respirators.

How Accidents Happen

Accidents often happen due to *inattention.* When using a plaster pump or sprayer, the pace is often set by the machine, and things can become frantic quite quickly. When a plaster job is rushed, the quality of the job will suffer, as can the safety of those present. Inspect all equipment regularly prior to use. Be particularly careful with plaster pumps; pressure can build up in the hoses that can release explosively — so all workers need to wear eye protection *at all times.*

Plaster is heavy. Make sure that buckets and wheelbarrows can be safely moved — there's no point in filling buckets to the brim if no one can lift them!

Sometimes, other crews are working on a jobsite at the same time as the plastering crew. If the site changes (e.g. if there is trenching going on, or heavy equipment is moving around the site), make sure that everyone on your crew is aware of the hazards.

Accidents regularly happen when ladders aren't set up on level ground, or when scaffolding isn't properly erected. Do regular inspections of all tools and ladders; make sure that scaffolding is set up according to regional safety standards. Wear a hard hat.

Be organized!

Falling, slipping, and tripping are among the most common jobsite injuries. One of the best ways to have an accident-free jobsite is to keep it organized. Plastering is inherently messy — sometimes it looks like a tornado has blown through at the end of a work day. Have a plan for regular cleanup, and have systems in place (i.e. set locations for where the buckets of plaster are to be kept vis à vis the plasterers, and a well-thought-out flow for equipment and materials, etc.).

Weather conditions

Extreme cold or heat can be problematic both for your plaster and for workers. Be aware of daily conditions, and be flexible about working hours (i.e. on hot humid days, you could start work early in the morning, and/or do a stint later in the day, when it's not so hot). Extreme winds can be dangerous to workers, as can extreme temperatures.

Using common sense

In an ideal world, there would be no accidents because everyone would use common sense at every step of the way on a jobsite. Reality tells us that, in fact, many accidents are preventable. One of the jobs of a site safety representative is to ensure that every reasonable precaution to prevent foreseeable dangers has been taken. It might seem obvious that using a power tool that creates sparks next to loose straw is a potential fire hazard, or that using a straw bale as a footing for scaffolding isn't safe, or that scaffolding that isn't securely tied to the building or properly leveled can tip, but we have seen all of these situations on jobsites. If we could rely fully on common sense, there wouldn't have to be construction safety requirements. Although some rules about safety on jobsites seem like overkill, they *must* be adhered to.

Personal Sustainability

Work habits can mean the difference between *damaging* your body and *strengthening* your body. If you want to continue working as a natural plasterer for a long time, pay attention to what your body is telling you and work hard to develop good work habits.

Lift with your legs, not with your back. It's an overused saying because it's true, and despite everyone knowing this phrase, few people consistently do it. Back injuries are common in construction, and they are often preventable — by lifting with your legs, and not your back.

Applying plaster by hand is a heavy job! If you are using a machine, it is fatiguing both physically and mentally if you factor in the sound of the equipment. Anyone who has been on a plastering site knows that sigh of relief at the end of the day when the mixer/plaster pump is shut down. Make sure to wear hearing protection around machinery; your hearing can't be regained once it's been damaged.

Repetitive stress injuries are common in construction. Plastering is no exception — a full day of plastering represents a high number of repeated actions: it could be thousands of strokes with the trowel, or lifting and moving large quantities of sand, binder, and water. Switching up jobs (when possible) can be helpful; if you are usually the mixer, perhaps part-way through the day, or the job, you might want to go and apply plaster, and have someone from the wall take over the mixer. Switching jobs gives everyone's muscles a chance to recover.

Finding more ergonomic ways to do certain tasks can help reduce strain on your body. Many tools are more suitable for larger hands, so if you have a smaller frame, try to find tools that are more comfortable for your body; they will fatigue you less. Stretching before and during the plastering day can be beneficial. Pace yourself, and, as a team, decide on realistic goals for the day. Olympic plastering days aren't really something to brag about — they're exhausting, and safety is jeopardized when people are overly tired.

Use warm water for cleaning if possible; it will be easier on your joints in the long run. Plan for adequate days off after big plastering jobs to allow yourself to recover. Long, hot soaks in the tub, massage, tai chi, yoga — all can be helpful.

Wear gloves and eye protection, even with earth plasters, but especially with lime-based plasters.

Safety Equipment and Training

There should be a fully stocked first aid kit on site in a central location; it should be well labeled and up to date. There should also be an on-site eye wash kit. A qualified first-aider should be present when plastering, and all personnel should be trained to use the eye-wash kit.

Fig. 1.4: *Although these plasterers may not be setting a fashion trend, they are well protected from lime burn.* CREDIT: LESLIE MCGRATH

Certain plasters are caustic, such as lime. Protective clothing, including long pants and shirts, should be worn, as should gloves. If using cement or lime, the gloves should be waterproof to protect against lime burn. In the plastering world, we often gush over new purchases of gloves, comparing their strengths, weaknesses — it's rarely about fashion; it's about function and durability.

If a caustic plaster gets on your skin, wash it *immediately* and rinse with vinegar. We have found that vinegar is a useful item in our kit, as it helps to neutralize the lime; it can be diluted with water to cut the sting. (Incidentally, vinegar can be a useful agent to clean cement or lime plaster off of wood [and in the rinse cycle of laundry to get lime off of clothing].) Minor burns are common despite protective clothing, so it is helpful to have something in your first aid kit to soothe a minor sore or burn (vitamin E gel capsules work well, as does aloe vera). Any cuts should be well protected from plaster.

Have emergency phone numbers, such as 911 (if applicable), hospitals, and fire department posted on site, as well as a map to the hospital. These will save time in the event of a true emergency. Have an emergency plan in place, so that all crew members will know what to do.

Make sure to keep a fully charged fire extinguisher on site and in a visible place.

The Law

Depending on where you live, there will be different national and regional construction safety laws and guidelines. Make sure you are familiar with all that apply to you and your project. Fines and repercussions can be significant. Construction site safety and the law is everyone's responsibility.

Chapter 2

Natural Plaster Ingredients

How Natural Plasters Work

PLASTERS ARE MADE UP OF FOUR MAIN COMPONENTS: binder, sand, fiber, and water. Many other ingredients can be added, and sometimes the fiber or the aggregate is left out. But these are the main ingredients that define a plaster; the most important is the binder.

Over thousands of years of natural plastering, four major binders have traditionally been used: clay, gypsum, lime, and cement. All are variable in their properties, so each has its own section in this chapter.

Ratios

The golden ratio of binder to sand is usually given as 1:3. This ratio usually works, but it is an oversimplification. In reality, the ratio depends on several factors, especially the type of sand you are using, the depth at which you will apply the plaster, and the end result you are seeking. Anywhere between 1:2 and 1:3 binder:sand is common.

The ratio of binder to sand is determined by the volume of the binder that is needed to fill the voids in the sand. With a well-graded sand (most commercial masonry sands) this ratio will probably be close to 1:3. However, the ratio will vary significantly depending on the sand, sometimes being as low as 1:2 — or even less. There's a simple way to test this. Take a sample of sand that has been dried in the oven, or has had prolonged drying in hot sun. Place a measured amount of it in a bucket and fill it with a measured amount of water until the water level exactly reaches the top of the sand. The amount of water you poured in is the volume needed to fill the voids in the sand, and the ratio of water to sand is also the ideal ratio of binder to sand.

Is this always the ratio you will use? By no means. You would not usually add less binder than this, because it could result in a significant weakening of your plaster. But you might add more binder than the ideal amount to fill voids. The result will probably be a harder, more durable plaster, but one that will be more prone to shrinkage cracking. More binder is often used in fine finish plasters because it results in a smoother and more polished plaster, and because these plasters are applied in thin coats and are less prone to developing shrinkage cracks. High binder is also used in some earth plaster base coats that have a lot of straw or other coarse fiber to control shrinkage cracking, because the extra strength is desirable. In this case, the fiber is acting as a partial substitute for aggregate.

Fig. 2.1: *All the ingredients for one mix of lime-stabilized earth plaster.* CREDIT: DEIRDRE MCGAHERN

Volume vs. weight

On the majority of jobsites, plaster measurements are made by volume. This is done for several pragmatic reasons: because of the relationship between voids and binder volume just discussed, because it's easier and faster, and because the weight of materials varies a lot depending on how wet they are. However, density also varies with water content, so volume isn't a perfect measure. Some plasterers work exclusively by weight, particularly artisans who specialize in fine finish plasters and mostly use dry ingredients. Weight can also be useful when small quantities of an ingredient are added and consistency is desired, such as with pigments. Recipes in this book use volume measurements; they will need to be adapted if you work by weight.

Table 2.1: Clay at a Glance

Uses	Interior plasters. Exterior plasters only in special circumstances.
Permeability	High (18 US perms)
Embodied energy	Low
Compatible binders	Lime, gypsum, cement.
Safety	Clay dust is very hazardous to breathe due to the presence of very fine silica. Always use proper protection during mixing and cleanup.
Key properties	High shrinkage. Can be reworked after drying. Very flexible (low structural cracking). Relatively easily damaged and repaired.

Fig. 2.2: *The adobe settlement in Taos Pueblo is estimated to be about 1,000 years old. The multi-story buildings are remudded as required and well maintained.* CREDIT: TINA THERRIEN

Introducing the Binders

Clay

Clay is usually considered to be the most ecological of all binders because it can simply be dug from the ground and used. Even when it is mined and sold industrially, the energy cost of processing it is typically lower than for other binders. Unlike other plasters that undergo an irreversible chemical set after being applied to the wall, clay (earth) plasters can be repeatedly wetted back to a workable state, and then dried again.

Earth plasters have a suite of distinctive properties, including high vapor permeability and flexibility, which make them ideal for use in natural building systems. However they are not weather resistant, so they are not generally suitable for exterior use as a finish plaster.

It's important to realize that one can't simply substitute one type of clay for another in a recipe and necessarily expect the same result — or even that the plaster will work properly.

History

Until very recently, earth was the most common building material in the world, and it is still widely used in Africa, parts of Central and South America, the Middle East, India, China, and Southeast Asia. In Europe, earth was used in most countries including France, Spain (adobe blocks), and Britain (cob). In the southwestern U.S., adobe was (and to some extent still is) a common building material, and sod homes were often built in the prairies — some are still lived in today. Many of these buildings were plastered with earth plasters, sometimes with lime, and these traditions inform many of our modern earth plastering practices. (In her book *Clay Culture,* Carole Crews delves into the long history of earth building in New Mexico. It's well worth reading.)

Origins and chemistry

> One might think of clay, then, as being almost a representative sample of the crust of the earth after it has been disintegrated and pulverized to very fine particle size by the action of erosion.
>
> — Daniel Rhodes, *Clay and Glazes for the Potter,* 1957

Clay is the product of many thousands of years of erosion of rocks (particularly feldspar), and the deposition of very fine particles, often on ancient lakebeds. Chemically, clay is primarily composed of the mineral *kaolinite* ($Al_2O_3 \cdot 2SiO_2 \cdot 2H_2O$), but with widely varying quantities of aluminates and silicates, as well as oxides of iron, calcium, magnesium, and many other compounds/impurities. But this doesn't tell us much about what clay actually is: incredibly fine particles usually flattened into miniature platelets. It is the interaction of these platelets that gives clay its properties.

The platelets are readily lubricated with a layer of water, and they have a small electrical charge that helps them stick together very strongly, yet they still slide over each other — making clay extremely plastic and malleable when wet, but quite hard when dry (because the lubricating layers are missing). The collective surface area of platelets is huge but also variable — a single gram of clay soil can have a total surface area of anywhere between 10 and 800 square meters depending on the type of clay.

As clay becomes saturated, it expands. As it approaches becoming fully saturated, it tends to resist further water penetration.

Dry clay maintains the shape it had while plastic; however, as the water disappears from between the platelets, the clay shrinks a lot. This is why, as a general rule, clays with greater plasticity and workability (and more platelets) tend to have higher shrinkage rates. It's also why, as with most binders, it's important to balance the amount of clay in a plaster with fiber and aggregate — to reduce shrinkage cracks. Clay plaster base coats tend to be very fiber rich compared to plasters made with other binders.

Types of clay

Site soil

Evaluating soil types

Site soils differ in the amount of clay, silt, and sand they contain. Ideally, soils used in earth plasters should contain 20–30% clay — or more. Soils with more than 30% clay can sometimes be substituted directly for pottery clay in recipes (but you should still test the resulting plaster). In fact, a good clay-rich site soil is often considered to make a stronger plaster than would pottery clay in a recipe. It's often possible to make a good plaster with as little as 10–20% clay content in your soil, but test — always test!

Silt can be either benign or harmful in plaster, depending on how much there is. Preferably, it would be less than ¼ of the clay content of your soil. This rule can bend a little, but if the amount of silt in the soil is equal to the amount of clay, it probably won't make a very good plaster. In general, you should aim to use the best soil possible for your plaster, which may mean trucking it in or using bagged clay. There are several ways to test your soil; you should do all of them and compare the results.

The ball test

The ball test is a first quick test, but don't depend on it solely, as it is imprecise and may give false positives. Moisten a handful of your soil and knead it until it feels consistent. You want it to be malleable and moist but not wet — roughly the consistency of playdough. Form it into a ball and drop it from shoulder height. If your soil has low clay content, it will tend to fragment; with higher clay content, the ball will simply flatten somewhat.

The ribbon test

Now take the damp soil and squeeze it between thumb and forefinger to produce a ribbon about ⅛ inch (2–3 mm) thick and less than ½ inch (about 1 cm) wide. Keep pushing the ribbon out to see how long you can extend it before it breaks. A minimum of 1½ inches (4 cm) usually indicates at least 20% clay. Evaluate the feel as you squeeze it: does it feel smooth and plastic or can you feel sand grains in it?

A variation on the ribbon test is the worm test — try to roll the soil into a worm shape and see how long and thin you can make it. The longer you can make it, the higher the clay content.

The jar test

The jar test can be a fairly accurate way to determine your soil type, but it takes time because clay can take days to settle out of suspension. Any jar will work for this test; a 1 liter mason jar is a nice size. Have a timer (with an alarm) and a permanent marker handy.

- Fill the jar no more than ⅓ full with soil, then top it up to ¾ full with water. Optionally, you can add some detergent or salt to help disperse the clay particles.
- Shake the jar well, until you feel that any soil clumps have broken up. If not using a dispersant, you may need to let it soak, and come back to shake it again. Start the timer when you stop shaking.
- After 40 seconds, all of the sand component of the soil will have settled out — mark this level on the jar.
- After about half an hour, most of the silt will have settled out, mark another line at this level.
- When the water is fairly clear (a day or more) the clay has settled out.

Fig. 2.3: *A jar test can tell you a lot about what's in your soil, but you often need a timer to be sure where layers begin and end.*
CREDIT: MICHAEL HENRY

Measure the height of each layer and divide it by the total height of the layers to obtain a percentage by volume of the soil. The measurement of clay will be high because the clay hasn't had time to settle. Clay continues to compact over time — even more so as it dries. To improve the accuracy of the test, the levels can be measured after the sample dries, but in practice we rarely do this.

The importance of tests

Plaster tests patches are essential when using site soil — unless you have used the identical soil (dug from the same spot), on the same substrate, in the past.

Trucking soil in

Maybe you want to use local clay, but there's none on your site. Does clay soil occur near your building site? It may be worth trucking it in. The clay itself will usually cost less than the shipping — for short hauls, the total price tag can be very reasonable, and this might be a better option than using a soil that isn't quite good enough.

Sports field clays

Infield mixes for ball diamonds are formulated very close to the needs of plasterers, though they sometimes have too much silt. A typical infield mix would be about 60–70% sand, with the remainder made up of silt and clay. Commonly, the silt fraction is about equal to or more than the clay fraction, but this is variable. Good infield mixes have a nice diversity of particle sizes in the sand, perfect for plaster.

Infield mixes are fairly universally available (because ball diamonds are ubiquitous) but shipping will vary, and you'll need to find out the ratio of sand:clay:silt, which may be as easy as a phone call. Arrange to get a sample before ordering a truckload. The goal with infield mix is usually to just add fiber and go — if a sand or clay delivery is necessary to modify the mix, it becomes less worthwhile. Once you've found a mix that works, your recipe can remain consistent, and you can have it delivered to any jobsite in the area.

Bagged clay

In some cases, bagged pottery clays are the most logical way to get clay. For veneer plasters over drywall, bagged clay is often a good way to go because it is uniform, free of contaminants, and available in a variety of colors. But bagged clay sometimes makes sense even for base coats if site clay is unavailable.

Bagged clay requires little or no testing once a recipe is established, and it is easy to estimate and mix for crews who are used to working with bagged product. Bagged or bulk clay may be available from a number of local sources; in many urban areas, the supply chain for selling bagged pottery clay in the relatively small quantities we need is well established.

Economics

Dry pottery clay ships in 50 lb bags. Once bulk discounts have been applied, and if the distributor is close to the building site, the price may be comparable to cement and lime products. Since bagged clay can be used directly in the mixer with no processing, the labor savings from using it vs. site clay might justify the cost of buying it, depending on shipping costs. This is less true if high-quality local clay is available, and, of course, bagged clay will have a higher embodied energy and ecological footprint than local clay will (but less than that for lime or, especially, cement).

Terminology

When choosing a pottery clay for your plaster, it helps to understand the vocabulary that potters use to describe clays. A good resource if you need more information about clays is the staff at your local pottery supply store.

Plasticity is one of the most important attributes of clay, and it is closely related to strength and shrinkage. Generally, clays with very fine particle sizes have high plasticity and high strength because the clay has a lot of binding power (which is good), but they also have higher rates of shrinkage (not so good). On the whole, plasticity is a great thing, and clays that lack plasticity (called *short clays*) are less desirable for plaster. *Water of plasticity* is the measurement used by potters for how plastic a clay is — the higher it is, the greater the plasticity. Dry bagged clay will improve in plasticity after being mixed with water; we find that letting the mixed plaster sit for a few hours can have a positive effect on workability.

Shrinkage increases with plasticity; however, some clays with only moderate to good plasticity have relatively high shrinkage — particularly kaolins. In evaluating clays, it's the *dry shrinkage* that matters, which typically ranges from 5% to more than 6%. (Dry shrinkage refers to the amount that a clay will shrink as it dries. If the percentage is too high, there will be far more shrinkage cracks.)

Some clays are considered to have more *tooth* (variation in particle size, which can reduce slumping and add dry strength). This results from the granules of the clay being less finely ground, and/or other impurities that may be present; it is largely a product of how the clay was processed. When a mesh size is given, it indicates a clay that has been processed with a hammer mill and screened; other clays may be *water washed* or *air floated.* The larger particles that are typically found in a screened clay give it more tooth, but may lower plasticity slightly. This size diversity may cause the clay to act more like site soil, and have a better result for a body coat.

Types of pottery clay

Many types of bagged pottery clay can be useful in earth plasters. In fact, a number of them can be used in either body coats or finish coats. The main exception is kaolin clays; most work well in thin finish coats, but not in body coats.

Ball

Ball clays have very fine grain size and, thus, high plasticity and shrinkage. They are the most plastic clays used by potters or natural plasterers. Ball clays are often stratified with coal seams, and their dark gray or brown color comes from organic matter in the clay. Ball clays are commonly used in base coats.

Kaolin

Kaolins, used in porcelain, are the whitest clays available to the plasterer. They are nearly pure kaolinite (clay mineral) with very low impurities. The particle size of kaolin clays tends to be

Table 2.2: Characteristics of Bagged Clay

Name	Clay Type	Plaster Type	Colour	% Silica	% Dry Shrinkage	Water of Plasticity	Notes
Redart	Earthenware	Body or finish	Earthy red	10–20			Limited plasticity. Moderate strength.
Hawthorn Bond	Fire	Body (or thick finish)	Light tan	0–50			Very good plasticity and workability, with a fine but variable particle size. Good "tooth."
Tile 6	Kaolin	Finish	Warm white	0.1–1			An air-floated plastic kaolin with relatively high green strength.
EPK	Kaolin	Finish	Grayish white	0.1–4	5.8	26	Edgar Plastic Kaolin. Plastic, fine-particle.
Kentucky Stone	Ball	Body (or thick finish)	Gray	10–30	5.4	26.6	Coarse-grained ball clay with good strength and plasticity. High silica content.
Kts-2	Ball	Body (or thick finish)	Gray	10–30	5.4	32.6	Intermediate-grained ball clay with very good plasticity and strength.
OM-4	Ball	Body (or thick finish)	Gray	10–30	6.2	32.8	Fine-grained ball clay with excellent plasticity and strength. Commonly available.
Bell Dark	Ball	Body (or thick finish)	Gray	10–30	6.1	33.5	Very plastic. Commonly available.

large — so, while they vary greatly in their properties, they are generally less plastic than many other clays. Compared to the other extreme, ball clays, the most plastic kaolin falls short of the least plastic ball clay.

Because of their purity, kaolin clays are low in free silica and tend to be the safest clays to work with. Kaolin clays are typically used in finish coats.

Fire Clays

The types of clay referred to as *fire clays* are a hodgepodge — the term *fire* simply refers to the ability of a clay to withstand high temperatures without melting. Fire clays vary amongst themselves in most properties, including plasticity, however they are typically not ground as finely as other clays, and so commonly have more *tooth*. When fire clays have high plasticity, they are good for base coats.

Bentonite

Bentonite is not for plaster. It has such fine particle size that it behaves quite differently than other clay. Bentonite is typically 10 times finer than ball clay. Although highly plastic, its very high rates of shrinkage and very low permeability make bentonite unusable in natural plasters. Potters sometimes mix small amounts (2–5%) of bentonite into clay bodies to add plasticity. It's possible that similar additions could benefit natural plasters, but there are many unknowns, including reductions in permeability.

Properties and uses

Clay, or earth, plasters are the most vapor permeable and flexible of all the natural plasters — they readily allow humidity to pass through, and they adapt to movements of the substrate without cracking. These properties are important when plastering over natural wall systems.

However earth plasters trade these virtues for lower impact and erosion resistance — earth plasters can erode relatively quickly under driving rain. Clay also has very high shrinkage as it dries, so earth plasters are either applied very thinly, or they contain large amounts of fiber and/or aggregate.

Clay tends to protect any natural material it is bonded to because clay is more *hydrophilic* (water loving) than wood, straw, etc. So earth plasters actually pull moisture out of adjoining materials, then let the moisture dry to the outside of the plaster in dryer weather. Clay is very permeable, letting moisture move through it readily, but it resists liquid water because, as it becomes wet, it swells and becomes less permeable (thus becoming hydro*phobic*). Earth plaster also tends to moderate relative humidity in the air, adsorbing moisture onto the clay particles when the air is above 50% relative humidity, and releasing it when humidity drops lower.

Clay can be blended with any of the other binders to modify the properties of each (see "Blending Binders" later in this chapter) as long as it is blended in the correct proportions.

Earth Plaster Coats

Earth plaster is usually applied in two or three coats. Unlike lime-based plasters, where the base coat and finish coat often have very similar recipes and properties, earth plasters often vary a great deal between coats. Over straw bales or a similarly soft substrate, a bonding coat of clay slip must be applied before the base coat. Earth base coat plasters (see Chapter 5) are applied much thicker than finish coats and contain a lot of coarse fiber. Earth finish coats (discussed in Chapter 6) sometimes contain fiber, but if they do, the fiber is usually much finer than what is used in base coats. Finish coats may be applied anywhere from $\frac{1}{32}$ to $\frac{1}{4}$ inch thick.

Letting Plasters Age

The Japanese have a rich history of creating perfectly detailed and beautiful earth-plastered structures, and they are also preeminent when it comes to *aging* earth plasters. In Japan, plaster is often mixed, complete with straw, and allowed to sit (or "brew") for a period of weeks or years. It is said that this increases strength, workability, and plasticity, and it reduces cracking.

In the North American context, it's hard to imagine planning this far ahead. In fact, when we first started working with clay, we were surprised to discover that some of our finish plasters needed to sit for a few hours or overnight to attain good workability. Now we know that most earth plasters will improve if left to sit for hours or days after mixing. This allows the clay to more fully hydrate and reach maximum plasticity. A beneficial process of fermentation begins and other positive reactions occur between straw and clay. That said, many earth plasters can be used almost immediately after mixing when need dictates.

Additives

Manure

Manure has two functions in earth plaster: it can add fine processed fiber, and it can add strength and water resistance. A variety of manures can be useful in plaster, but the most commonly used are horse and cow.

Horse manure is good for adding fine fiber to the plaster, whereas cow manure contributes more enzymes that add strength and waterproofing. Cow manure has a much stronger smell — but don't panic, the smell goes away.

Manure should be fairly fresh and not composted, which destroys enzymes and fibers. If the manure is lumpy, you may need to push it through a fairly fine screen (⅛" or even ¼" works). You can either sieve it dry/damp, or blend it with some of your mix water to make a soup, and sieve that. Sieving fairly fresh cow manure has the added advantage that it may kill some of the larvae that are in it, which otherwise leave trails and exit holes from your plaster.

We usually blend manure with mix water using a paddle mixer because the resulting mix is faster to sieve. Dryer manure is harder to sieve, but will have less of a smell.

Lime

Adding lime to clay creates *lime-stabilized earth,* which is a plaster with properties that differ from either earth or lime. It is important to use the appropriate amount of lime for this reaction to work. Lime is covered in some detail later in this chapter, and Chapter 7 is devoted entirely to the subject of lime plasters.

Starch Pastes

Starch pastes glue the particles of plaster together, resulting in very hard plasters that resist many kinds of abrasion and have little or no dusting. Starch pastes are especially useful for interior finish plasters that won't be painted. In such cases, starch paste, or something similar (such as casein) is usually used to prevent dusting.

Wheat paste and rice paste have similar properties, but they require cooking, whereas pre-gelatinized starch comes in the form of an instant powder. When starch pastes are used, they usually make up 5–10% of a recipe by volume (excluding water), but the amount can be much higher in the case of *alis* (clay paint) or certain finish plasters. Five percent adds some strength, and the plaster is still very workable. Ten percent or more makes a very strong plaster, but it can be sticky and difficult to apply. There's generally a learning curve to working with plasters that have over 10% wheat paste in the recipe. A rest period of at least a few hours (and up to overnight) between mixing and application will make the recipe more workable. However, plasters with wheat or rice paste should be used within a day or two of mixing, especially in warm weather — otherwise, they will begin to decompose, mold, and smell unpleasant.

Pre-gelatinized starch is wheat starch that has been pre-treated so that it forms wheat paste with the addition of cold water, no cooking needed, and it has several big advantages. It's a purified form of the starch, and therefore has less unwanted food in it that could promote mold growth or lead to bad smells. It can be used immediately without processing; the labor cost to make wheat paste is probably greater than the cost to buy pre-gelatinized starch. A downside seems to be that it may be more prone to leaving unsightly drying marks on the plaster surface, particularly when the plaster dries very unevenly. Pre-gelatinized starch can be difficult to source — it's used as glue in the restoration of old books — online retailers specializing in conservation supplies may stock it, or some wholesale food suppliers.

Casein

Casein, a milk protein, is the main binder in milk paint. It is mostly used in natural paints, but may be used in finish plaster coats. Sometimes plasterers will throw a variety of milk products into all sorts of earth plasters, a sort of cowboy approach to introducing casein.

Borax

Borax is sometimes used as an additive in plasters, the idea being that it reduces the likelihood of mold on the plaster surface, especially when drying conditions are poor, but it is generally used in very small amounts.

Oils

Oils, usually linseed oil, can be soaked into the surface of plasters to create a durable, waterproof surface identical to an oiled earth floor. Several heavy applications of oil may be needed to do this. However, the permeability of the plaster will be reduced dramatically; so, if moisture does get behind the plaster, it could cause delamination. Alternately, we've heard of people adding less than 1% oil by volume of the mix, but we've never tried it. An addition of some amount of oil may increase toughness and flexibility, but there is still the same risk of delamination from the other earth plaster coats an oiled plaster is bonded to.

Fermented Products

Among the leading proponents of fermented products in plasters is the French builder, Tom Rijven. In fact, fermentation is key to his body coat recipes. Rijven adds fermentable products to base coat plasters — typically replacing half the straw with highly fermentable grass clippings — then he adds a fermentation liquid (a starter culture) that comes from a fermentation vat. The fermentation liquid may come from submerged corn silage or some other sugar-rich base. The plaster is aged for a few days in a warm, low-oxygen environment, then applied to the wall. Fermentation continues until the plaster dries on the wall.

Tom has found that when a plaster with grass ferments, bacteria feed on the glucose present in the fiber, elongating the fibers and resulting in more reinforcement. In addition, he finds that fermented plasters are more durable and waterproof.

A more common way to add fermented/fermentable material to the wall is to use manure, particularly cow manure. Whichever approach is used to introduce fermentation, expect bad smells. However, they will go away after the plaster is fully dry.

Using straw as a fiber in earth plasters starts a chemical reaction that adds strength and waterproofing to the plaster. In his book *Earth Render,* James Henderson describes tests by Allen Kong who found that simply boiling straw and using the water strained off of it added significant strength and hardness to mud bricks.

While a minority of plasterers intentionally add fermented products (other than manure) to their plasters, many a wall has benefited from the fermentation that occurs naturally when earth plasters containing straw or other organics are left to age before being applied.

Safety and Handling

Clay contains very fine silica in widely varying amounts, from less than 1% to greater than 50%. The particles are so fine, it is easy for them to become airborne. Inhaled silica causes chronic debilitating disease and death, so wearing a proper respirator during mixing and cleanup is essential (see "Choosing respirators and vacuum filters," in Chapter 1).

Kaolin clays often contain less than 1% silica, making them good for earth plaster finish coats. Ball clays and fire clays are more common in earth plaster base coats, and typically have large amounts of free silica. Site clays are typically processed wet, so they are hazardous only during cleanup.

Lime

Lime is used all over the world for foundations, walls, plasters, mortars, and for decorative cornices and moldings. It can be used in a wide variety of applications, including exterior plasters, and can have many different finishes — from pebble dash to smooth and highly polished. Lime plaster has stood the test of time; it is permeable, yet water resistant, has fungicidal properties, and is relatively inexpensive.

Table 2.3: Lime at a Glance

Uses	Interior and exterior plasters, cornices and moldings.
Permeability	Good (14 US perms)
Embodied energy	Medium
Compatible binders	Clay, cement, gypsum
Key properties	Is self-healing (can be wetted down and will fill in cracks while in the curing stage). Requires well-graded sharp aggregate. Relatively slow to cure. Must be applied in subsequent thin coats. Fungicidal properties.

History

Lime has been used in construction for at least 9,000 years. The earliest known uses of burnt lime is in floors and plasters in the Middle East; it was widely used in Greek and Roman architecture. The use of lime-based plasters is evident all around the world. You can find it in the Pantheon of Rome, Michelangelo's frescos, and in the ruins of Mexico and Peru.

Lime plaster in Britain in the 1200s was used for structural and fireproofing purposes, and decorative elements were adopted later. In 1501, King Henry VII granted a charter to the Worshipful Company of Plaisterer's in London. In the United States, plaster guilds formed in Philadelphia in the 1790s. The long line of the craft of plastering was thus passed on to journey apprentices in the guild.

Apprenticeship programs and guilds are emerging once again today. If you have the opportunity to work alongside someone from a lime plastering tradition, jump at the opportunity to learn from them. This knowledge will be lost if it isn't passed on. You may have to travel far to find an expert, but it will be worth the journey.

Origins and chemistry

Lime is manufactured from limestone, which is sedimentary stone created from the skeletal remains of marine organisms, layered with clay and silt — simply put, it comes from seashells and the skeletons of plankton accumulating, compacting, and eventually hardening into rock that contains high levels of calcium carbonate. There are different kinds of limestone — some contain clay, aluminum, iron, or potassium, others have magnesium, and some are relatively pure calcium carbonate. Limestone with large amounts of magnesium is referred to as dolomitic limestone, which is common in the U.S.

Impurities in limestone can dramatically change the properties of lime used in plastering. These changes can sometimes be quite useful because they create *hydraulic limes* and *natural cement* — but more on that later.

Limestone, or calcium carbonate ($CaCO_3$), is taken from a quarry, crushed, washed, and then heated to 1500°F (900°C). The heat breaks the chemical bond between the calcium oxide and carbon dioxide, resulting in the loss of carbon dioxide (CO_2) and leaving behind calcium oxide (CaO), commonly called *quicklime.*

Quicklime is a highly reactive form of lime. It can be purchased in the form of pellets, lumps, or sometimes powder; care must be taken to wear protective gear when working with it.

When water is added to quicklime, heat is given off, and the resulting product is *hydrated lime* (calcium hydroxide, $Ca[OH]_2$). In North America, manufacturers of lime add precisely measured amounts of water (in the form of steam) to the quicklime under pressure. This transforms the quicklime into *dry hydrated lime* which can be bagged and shipped out for immediate use in plasters and mortars.

If the $Ca(OH)_2$, or hydrated lime, continues on in the lime cycle, and is allowed to soak, or *slake,* in more water, it forms what is called *lime putty,* which must usually mature before it can be used (although North American Type S limes do not require extended slaking to be useable).

When lime in any of these three forms is mixed with sand and water (and often fiber) to form a plaster, it absorbs CO_2 from the atmosphere as it cures and hardens.

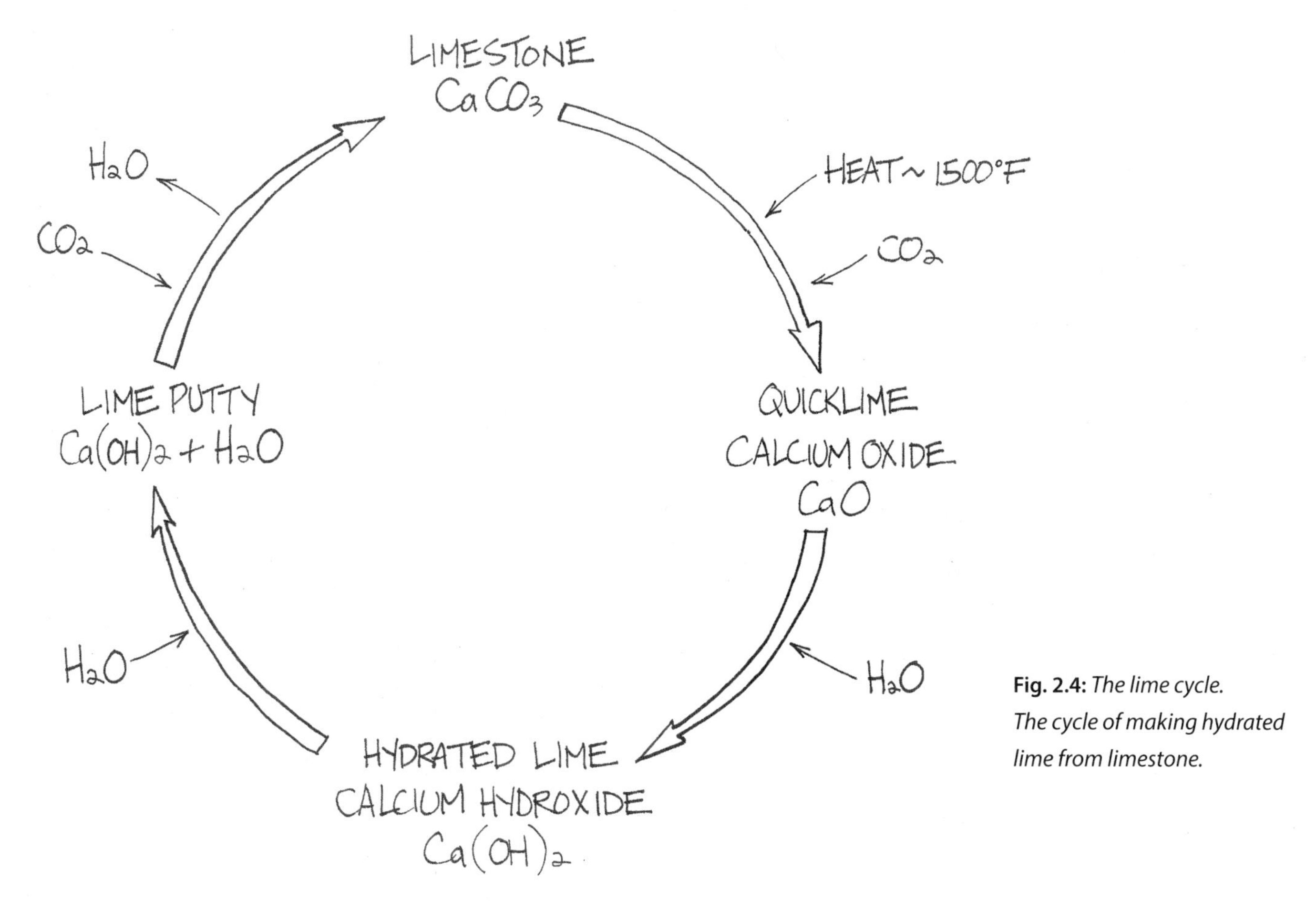

Fig. 2.4: *The lime cycle. The cycle of making hydrated lime from limestone.*

And thus we have come (sort of) full circle, with the resulting calcium carbonate being very similar in chemical composition to the original limestone. Hydraulic lime undergoes the same process, but has a slightly different cycle due to its impurities. Reactive silicates are a by-product of this cycle after water is added to hydraulic quicklime, so technically, it isn't a closed cycle (it's non-repeatable).

Manufacture

To obtain lime, limestone is heated to high temperatures in a kiln. Modern kilns are fueled by gas or coal, but the earliest kilns were merely dug out of hills, lined with heat-resistant rocks, and fueled by wood. Kilns in subsequent years, such as draw kilns or stack kilns, were more sophisticated, with thick stone walls and a tall chimney for draft. The ash from the wood fire would sometimes mix into the lime, which wasn't necessarily undesirable in the plaster, as it created a *pozzolanic effect* on the plaster (see "Pozzolans," later in this chapter). By the end of the 1920s, the age of draw kilns had come to an end because modern gas-fired kilns made the process more efficient. But historic quarries, lime kilns, and slaking pits can still be found around the world close to where limestone deposits are located; they dot the countryside in North America, having been attractive sites for settlers.

While there is pollution emitted during the manufacture of lime, it gains back some environmental points by reabsorbing some CO_2. When 100 lbs of limestone is kilned, it yields 56 lbs of quicklime and 44 lbs of CO_2. Additional CO_2 is released when fossil fuels are burned to heat the kiln. When lime plaster carbonates, it reabsorbs much of the chemically released CO_2, but the CO_2 that was released in the burning of fossil fuels is outside this cycle and remains in the atmosphere. Depending on the impurities in

Fig. 2.5: *Wood-fired draw kilns dotted the countryside a century ago. It was typical to have slaking pits near the kilns, wherein the quicklime was slaked with water and buried for at least a few months. In some places, the lime was slaked for a year or more.*
CREDIT: MICHAEL HENRY

the lime, the amount of CO_2 that is reabsorbed will vary, with the result that hydraulic limes reabsorb less than other types of lime. Portland cement does not reabsorb CO_2 the same way, and in fact when it is blended with lime, it blocks effective carbonation and absorption of CO_2 by the lime.

Types of lime

All types of lime plaster can be tricky to work with — there is a definite learning curve associated with them, and each type has its own set of challenges. It likes weather that is "not too hot, not too cold, but just right." It needs to be protected from sun and wind, and regularly misted after application — for a week or more. Types of lime to consider using include natural hydraulic lime, dry hydrated lime, lime putty, and quicklime. Each has its place, and one important factor to consider is whether you can obtain the material locally or not. In North America, dry hydrated lime is the most readily available. In our region (Ontario), natural hydraulic lime is also available, although it must be imported.

Natural hydraulic lime (NHL)

When limestone that contains impurities such as clay or amorphous silica is burned to create lime, natural hydraulic lime (NHL) may be created. (In the U.S., you may see the term *hydrated hydraulic lime* [HHL]; it is the same material as natural hydraulic limes.) Hydraulic limes behave quite differently than non-hydraulic hydrated limes. The impurities create a different set of chemical reactions that give a hydraulic *set*, meaning it will start to set as soon as water is added, even in the absence of air (hydrated lime, by contrast, can be mixed into a putty that will store indefinitely if it is covered with water). The advantage of the set is that natural hydraulic lime cures in days instead of weeks, and the resulting plaster is a little harder and less porous. Due to the faster set time, NHL may be a better choice than hydrated lime for plastering close to frost season, but it must be protected from frost until fully dried, and it will continue to cure and gain strength for months. Hydraulic lime is less permeable than hydrated lime.

When talking about NHL plasters, there are two standards, the European Norm EN-459, and the American Standard ASTM C-141. Note, though, that any lime that has had a pozzolan or other material added to it, either in the kiln or afterward, is referred to as an *artificial hydraulic lime* (AHL), and it doesn't conform to either standard.

In Europe, there are three main strengths of natural hydraulic lime that are available:

NHL 2	Contains 6–12% reactive clay; feebly hydraulic (softer)
NHL 3	Contains 12–18% reactive clay; moderately hydraulic (medium hard)
NHL 5	Contains 18–25% reactive clay; eminently hydraulic (hard)

When trying to select the proper NHL plaster, keep in mind that the plaster should be of similar strength to the substrate. NHL 5 would be most appropriate over top of a solid masonry wall, for instance, whereas NHL 2 is a good choice on earth bricks or straw bale buildings.

Hydraulic limes are not as sticky or as readily workable as non-hydraulic limes (they feel sandy), but they do tend to crack less, as there is less shrinkage (the sand and lime fuse together more tightly than in non-hydraulic plasters). Natural hydraulic lime can't be used to make lime putty, as it begins to set with the addition of water.

Dry hydrated lime powder

Hydrated lime doesn't set in the presence of water alone, but rather, by the carbonation of calcium hydroxide to calcium carbonate via a slow reaction with atmospheric carbon dioxide while the plaster is moist. This plaster should remain damp for a week for the initial set, and should be rewetted periodically for several weeks after application in order for full carbonation to occur. Plasterers in North America use bagged hydrated lime purchased from masonry supply stores; unlike the type available in Europe, these bags contain high-quality lime, comparable to lime putty.

Hydrated lime is obtained when quicklime has a small amount of water or steam added to it during manufacture. Dry hydrated lime is sold in bags, and must be used when still fresh, because over time it reacts with air (in the presence of humidity), which will be evident if there are chunks in the bag. It's best to check the date it was bagged; aim to get lime that is less than a year old. Hydrated lime (including lime putty) needs to be applied in relatively thin layers; generally to a maximum of ⅜ inch (10 mm), so it may take three or more layers to level a wall.

Within hydrated lime powders, there are two types, according to ASTM standards: Type S (Special) and Type N (Normal). Within the construction industry, Type S is used almost exclusively, especially for plastering. Type N lime is produced with normal hydration (at atmospheric pressure), and generally contains a higher amount of unhydrated oxides. A Type N lime needs to be slaked in water to be acceptable as a plaster, but Type S can be used directly, as it has been adequately slaked in the factory. You may find books and websites that tell you *not* to use bagged hydrated lime for plastering — this reflects the reality in the UK and continental Europe, where high-quality Type S lime is far less available (and lime putty is far more common).

Type S gets its name from ASTM C207 standards; it refers to dolomitic lime (lime with magnesium) that gets a pressurized hydration in an autoclave, resulting in full hydration of both the magnesium oxide and the calcium oxide.

Agricultural lime is finely pulverized chalk (calcium carbonate) or limestone that hasn't been heated and chemically changed to calcium hydroxide. It is useful as a soil additive, but has little to no binding ability, and should not be used for plastering.

Hydrated lime putty

In the lime cycle, after steam is added to quicklime, it becomes dry hydrated lime powder. If that same hydrated lime is soaked (slaked) in water, it will become lime putty — a creamy, luxurious form of lime for plastering. The longer the lime putty has been slaked, the better. A minimum of three months is recommended for Type N limes. Lime putty can keep indefinitely, as long as it has a skim of water over its surface in an airtight container.

Lime putty is more readily available in Europe, where cement hasn't eradicated the use of lime. Some plasterers claim that by slaking even Type S hydrated lime, you get a much more workable, creamy plaster. We don't usually bother.

If you find a recipe for lime putty, but you only have dry hydrated lime, or vice versa, it is helpful to know that lime putty can have 1–1.5 times as much lime by volume as dry hydrate. So, if a recipe calls for lime putty, and you don't have any, you can substitute with Type S hydrated lime powder, but you may need to multiply the volume of lime by up to 1.5. Do some tests with the resulting plaster to make sure it performs properly before using it on an entire building.

When to use hydraulic lime vs. hydrated lime

Hydrated lime and lime putty make soft, flexible plasters that are suited to flexible substrates such as cob, straw bales, clay/straw, etc. This type of

Hot Lime

By Nigel Copsey

Hot lime is experiencing a resurgence in the UK and France. Until at least 1800 in the UK, and until later elsewhere, plastering systems included both earth-lime mortars and lime mortars. Until 1800, the majority of stone buildings were built with earth or earth-lime bedding mortars, pointed to the exterior with lime-rich, hot-mixed mortar and plastered within with an earth-lime basecoat over which was laid a haired, lime-rich finish coat of between 3⁄16 and 3⁄8 inches (5–10 mm). In our observation, similar systems predominate in Spain, Italy, France, and Ireland and were doubtless as common elsewhere. Ukrainian migrants carried the routine use of similar mortars into northern Alberta during the later 19th century.

There are different methods of hot mixing (described in Chapter 7). In our experience, mixing the lime and the aggregate whilst the lime remains hot delivers the best mortar. Quicklime was usually slaked to a dry hydrate when it needed to be transported long distances, or by sea, and this became more common for plastering during the 19th century. It was also not uncommon in Italy for fine stucco finishes, mixed with marble dust.

Contrary to common assumption, the use of quicklime is no more hazardous than the use of other routinely used alkaline binders, such as Portland cement or hydraulic or hydrated lime — if done correctly, with adequate knowledge and reasonable precautions. Properly slaked, the temperature of a hot-mixed mortar will not exceed 248°F (120°C) during slaking and will fall to between 122–140°F (50–60°C) once sufficient water to produce a workable mortar is added. This process takes a matter of minutes.

Quicklime is available in most parts of the world, and can be made on a small scale wherever there is a supply of suitable limestone or sea-shell.

lime is more porous, and thus more permeable, than hydraulic lime, and it works well on flexible substrates that may be apt to slight movement. It is an ideal plaster for interior finishes; when used on exterior walls, it must be paired with a suitable finish, such as a silicate paint. Hydraulic lime could fare better in exterior situations that are exposed to extreme weather, assuming the somewhat reduced flexibility and vapor permeability is acceptable. Using a weaker hydraulic lime, such as NHL 2, can be a good compromise.

There are significant differences in application season between NHL and hydrated lime, which can be important to plasterers. Hydrated lime is a good choice for early fall plaster, when weather is cool but risk of frost is a minimum of several weeks away. Hydraulic lime sets much faster and may be better suited to short windows of good weather in later fall (but it must not freeze while still wet, during the initial cure). If ice crystals get into the pores of a plaster, it will likely fail. Lime plasters that freeze before curing won't fully carbonate, nor will they develop full strength. This can result in crumbling or flaking of the plaster.

Hydraulic lime plaster is also more forgiving than hydrated lime during summer months, though hot windy days are still a no-go.

In North America, hydraulic limes are imported from Europe. Until recently, the price has been a significant barrier, but lately it has been coming down significantly, with availability varying regionally. The least expensive way to obtain the qualities of a hydraulic lime is often to take a hydrated lime plaster and add a pozzolan to it to achieve similar effects (see "Pozzolans," below).

Lime: A Summary

A well-balanced binder, lime is used for its weather resistance, permeability, and flexibility. Its permeability is less than that of clay, but it is still appropriate for natural buildings, and it works well with other binders. Lime plasters, which are blended with 1–3 parts sand, are extremely sticky, and generally they adhere well to most prepared surfaces, including metal or wooden lath, bale walls, or solid walls that have been primed appropriately (See Chapter 3). Relatively strong, yet flexible enough to move with buildings, lime doesn't crack in the same way that cement does. Lime plasters are autogenous (self-healing), meaning that when exposed to CO_2 and water, the uncarbonated, or *free lime,* can help fill in any cracks that form. Vapor permeable paints or other sealants can be important, especially on fairly exposed sites.

Pozzolans

Pozzolans are materials that enable plaster to set more rapidly. The word *pozzolan* is derived from a type of volcanic rock found in Pozzuoli (Naples). Pozzolans have been used in plasters for thousands of years. When pozzolanic materials are added to non-hydraulic limes, they react with the calcium hydroxide in the lime, resulting in a more cementitious plaster, similar to plasters used in the time of the Romans. Pozzolans serve two purposes: not only do they speed the setting time of the plaster, they also increase durability. In the 18th and 19th centuries, experimentation with pozzolans created hydraulic cements out of hydrated limes, and eventually led to the discovery of Portland cement. All lime plasters, whether hydrated or hydraulic, are altered by the addition of pozzolans — but they are most commonly added to hydrated non-hydraulic limes. Most pozzolanic materials are comprised of silica and alumina, along with clays and iron oxides.

There are natural and artificial pozzolans. Natural pozzolans are pozzolans that haven't had any artificial heat added to them, such as volcanic rock or ash. Diatomaceous earths and high-silica rocks are also natural pozzolans. Artificial pozzolans have had some external source of artificial heat added to them; these include lightly fired clay products (like tiles or bricks), blast-furnace slag, clays, tile and pot shards, burnt clays, forge scale, and wood ash.

Pozzolans suitable for use in plasters include the following:

- Volcanic ash.
- Lightly burnt clays (kaolinite).
- Welding slag.
- Metakaolin (manufactured from kaolin clay).

Table 2.4: Lime Properties and Uses

Hydrated lime	$Ca(OH)_2$	Interior or exterior, if well protected.	Slow set time; cures with exposure to air; bags of hydrated lime should be fresh; should be applied between 5°C–30°C and protected from freezing for several weeks.
Hydrated lime putty	$Ca(OH)_2$	Interior or exterior, if well protected.	Slow set time; cures with exposure with air; lime putty can last indefinitely as long as it is kept covered with water; should be applied between 5°C–30°C and protected from freezing for several weeks.
Natural hydraulic lime (NHL)	$Ca(OH)_2$	Interior or exterior use.	Fast set time; cures with exposure to water; more tolerant to cold weather application, but can't freeze until initial curing/drying is complete.
Quicklime (hot lime)	CaO	Interior or exterior use.	Extremely hot while mixing; can be used hot or cool; historic plaster.

- Ground brick dust (only if bricks have been fired at a low temperature — most modern bricks are fired at temperatures that are too high to create pozzolanic effects).
- Fly ash (a by-product of most coal-fired power plants in North America).
- Forge scales and ashes.
- Agricultural waste products including wood ash, rice husk ash, bagasse (sugar cane husks), and rice straw.

If brick dust is used as a pozzolan, it should either be mixed with water or wetted down prior to mixing into the plaster. In fact, most pozzolans will behave better if they are mixed with a minimal amount of water before being added to the plaster. Pozzolans should only be mixed into the plaster right before it is to be used, as the pozzolanic effects will start taking place immediately. A mix that is older than 24 hours should probably not be used.

The amount of pozzolan required for a plaster will vary depending on how reactive that particular pozzolan is. Higher pozzolan ratios will speed up the cure, but may affect the overall strength of the plaster. Holmes and Wingate (*Building with Lime,* 2003) state that an appropriate pozzolan ratio in a lime plaster is 1 lime:2 pozzolan:9 sand. Some historic mixes included ratios of 1–2 lime:1 pozzolan:1 sand. The exact ratio depends on the particular pozzolan in question. As always, make sure to do test mixes to find the right ratio for your mix. The Endeavour Centre shares a recipe for a lime-metakaolin base-coat mix in Chapter 7; metakaolin is a pozzolan.

Additives

Fibers in plaster allow for slight movement of the plaster as a building moves — without allowing cracks to develop. Fibers also reduce plaster shrinkage, which also helps ward off cracks in the plaster. Traditional lime plasters on lath would have contained hair (ox hair was preferred, but hair from horse, goat, or donkey have been used). Any hair to be used in plaster needs to be grease-free, strong, and in the range of 1–3 inches long. Human hair is not suitable because it is too fine and not particularly strong, although it has certainly been used. Other fibers include chopped straw, reed, manila, jute, and sawdust. Modern plasters may include synthetic fibers, such as polypropylene.

Other additives

Various materials have been added to lime plasters over the centuries, including ox blood, nopal cactus gel, egg, linseed oil, urine, seaweed, hemp, gypsum, molasses, casein powder, and manure. Some additives can assist with water repellency, reduce shrinkage cracks, and increase bond and strength, while others can act as *air entrainers* (air entrainers add tiny bubbles into plaster, using a surfactant, to give more protection against freezing). Each additive has its own purpose in a plaster, and tests should be done to determine their effectiveness in your plaster. Finding someone who has used a particular ingredient in their lime plaster can be helpful in determining quantities to use in the mix, and the expected results.

Substrate

Lime plaster can be applied over a variety of substrates, including, but not limited to, the following: drywall, wood lath, straw bales, and masonry walls. If lime is desired as a finish coat on a cob or other earth wall, there should be a lime-enriched clay base coat first to allow the two to bond.

Safety and Handling

Lime is caustic in nature, and can cause nasty burns on the skin. Proper protective gear is essential when plastering with lime; make sure you wear long sleeves, pants, rubber gloves, and safety glasses. If you are mixing lime, as with any plaster, a proper respirator should be used.

Natural and Artificial Cements

Finally, there's *cement* — which is a dirty word in natural building circles. The manufacture of cement contributes enormous amounts of greenhouse gases, which seems like an odd choice in a natural building. In some parts of North America, during the revival of straw bale building, cement became a go-to plaster for many bale builds. In part it was due to accessibility, and in part, due to lack of experience or awareness of other binders.

There are natural and artificial cements. Natural cement occurs as a result of very specific (aluminate) impurities in limestone. This argillaceous (clayey) limestone, when calcined, produces a hydraulic cement. Some pozzolans, such as metakaolin, also create weak cementitious reactions. On the other hand, Portland cement is an artificially created cement. Natural cement can be used almost interchangeably with Portland, except that it has a very quick set time (which can be partially managed using retardants). Portland cement has a higher embodied energy, and it contains more toxins than natural cement, but it is nevertheless more widely used in plastering natural buildings than natural cement because of its low cost, availability, and controlled set time. When we talk about cement, we therefore are usually referring to Portland, even though it is manufactured in a way that excludes it from being a "natural" plaster.

Cement plasters are very strong, but they are prone to cracking and have low permeability. Pure Portland cement has such low permeability that it virtually guarantees rot (when bonded to natural materials) — but when mixed at least 1:1 with lime to make a cement-lime binder, it does have some limited applications in natural building — if treated with caution. We include it in this book because we recognize that it is, and will continue to be, used in natural building. When it *is* used, we'd prefer it be used correctly.

Table 2.5: Natural and Artificial Cements at a Glance

Uses	Interior and exterior plasters
Permeability	Low (2–10 US perms)
Embodied energy	High
Compatible binders	Lime
Key properties	A very hard, strong and brittle plaster. Requires sharp well-graded aggregate. Relatively fast to cure. Low permeability limits acceptable uses. Prone to cracking. Controlled set times. Rapid strength gain.

History

In the 17th and 18th centuries, pozzolans made their way from Italy to England, where plasterers incorporated them into hydraulic limes used for bridge construction. When John Smeaton was engaged to build a new lighthouse on Eddystone Rocks in England in 1756, he experimented with various limes and pozzolans, finally discovering that limestone deposits containing significant amounts of clay, when calcined (burned), were hydraulic. It was Blue Lias lime, containing clayey marl, along with a pozzolan from Italy, that was used for the famous lighthouse. It is said that Smeaton opened the door to natural hydraulic limes and natural cements.

Experiments in France, England, and the U.S. led to the discovery of natural cements, which can be made by burning clay-rich limestone; the result is a fast-setting cement. One such cement was patented by James Parker in 1796. He called his product *Roman cement* (although it differed from the plaster that Romans would have used), and it was used extensively in the U.S. for bridges, forts, dams, and the Erie Canal due to its strength and water resistance. Natural cements, although similar in content to hydraulic limes, are very fast setting — sometimes there are only minutes to work with the product. The quick set is due to a high active clay content. The silicas and aluminates contained in the clay portion of natural cements are the components that cause the almost immediate set.

Natural cement production in the U.S. became well known around the world; this is partly because of the extensive limestone deposits in Louisville, Kentucky, and in Rosenberg, Texas. There were even a couple of natural cement factories in Canada at the turn of the century. Natural cement was used in North America from the late 1800s until the early 1900s. Natural cement can be seen as a bridge between lime plasters and modern cement.

Joseph Aspdin was an American brick layer and inventor. In 1824, he patented what he would call Portland cement, which was a combination of pure limestone, clay, and other minerals, such as silicates. Portland cement was named thus because it resembled the color of the limestone quarried on the Isle of Portland in the English Channel. These minerals resemble the chemical composition of marl (which is unconsolidated clay and lime rock, or soil), but are controlled in a laboratory setting. Portland cement proved to be more reliable than natural cement in that the mix was consistent every time. It allowed a longer working time, and it was strong. The manufacture of Portland cement didn't start in the U.S. until 1871 and in Canada in 1889.

On its own, Portland cement wasn't particularly workable, but with the addition of lime, it provided a strong plaster with good workability. The mix quickly replaced natural cements, and it became the most popular artificial cement. The Hoover Dam and the Grand Coulee Dam, built in 1936 and 1942 respectively, were feats of engineering that were made possible by Portland cement. Close to 12 million cubic yards of concrete went into the construction of the Grand Coulee Dam.

Origins and chemistry

Cement is made from limestone, plus clays and other minerals such as silica and alumina, all in particular ratios. They must be baked at much higher temperatures (2732°F, or 1500°C) than lime is; and they have to be baked for an extended period of time. Because of the large energy inputs needed for this amount of sustained high heat, a tremendous amount of CO_2 is expelled during the fabrication of Portland cement, contributing greatly to greenhouse gases. In the kiln, though, when CO_2 is given off, calcium oxide forms. The alumino silicates from the clay react chemically with the calcium oxide. Cement clinkers form in the high heat, which are then cooled and pulverized. As they are cooling, a limited amount of gypsum powder is added, forming what we know as Portland cement. While Portland cement is the most readily available cement in North America, natural cements are still

The Cement-Lime Controversy

Cement quickly replaced lime in many plaster applications in the early 1900s in North America. Cement became the new go-to plaster, and it was used to "protect" adobe buildings, brick homes, and plastered buildings. Formerly permeable wall systems became impermeable. Within decades though — and too late for some buildings — it became clear that the low permeability of Portland-based plasters was trapping moisture within walls, causing failures of the plaster or the substrates — or both. In some cases, buildings that had stood for centuries were irrevocably lost after "renovations" using Portland-based plaster.

Cement-lime plasters can be — and have been — used on straw bale buildings (including by the authors); however, we believe that they are never the best choice, and we advise against using them. Hopefully, there will continue to be a return to traditional earth and lime plasters when appropriate.

being produced, and they are more appropriate than Portland cement in restoration of construction originally done with natural cement.

Types of cement

There are different types of bagged Portland cement products. Some types are 100% Portland, while others have some lime content. When we have had to use cement for any of our plastering, especially on a bale building, we would typically seek out Type N Portland-lime, a blend of 50% Portland and 50% lime, which is to be mixed 1:3 with sand. In some of our later plastering, we used site-mixed Type O Portland-lime (1 part Portland:2 parts lime:9 parts sand). All Portland-lime blends are defined by volume, not by weight.

Safety and handling

As with the other binders, make sure to have an MSDS (material safety data sheet) on hand for cement. Portland cement contains crystalline silica, which can cause silicosis and other respiratory and autoimmune disease with prolonged exposure. Ensure that the people who are mixing wear proper respirators, safety goggles, sleeves, and gloves; the crew cleaning up dry plaster afterward also needs the same protective gear.

Gypsum

Gypsum is one of the oldest plasters, and, because it can be cooked as low as 350°F (177°C) to create a binder, it is among the most ecological. Most gypsum plasters can't be used in exterior applications or anywhere they will be exposed to moisture.

Table 2.6: Gypsum at a Glance

Uses	Interior plasters, moldings
Permeability	High (18 US perms)
Embodied energy	Low-medium
Compatible binders	Clay, lime
Key properties	Expands while curing (no shrinkage cracking). Does not require aggregate. Sets quickly. Good depth tolerance. Fire retardant. No weather resistance.

History

The best preserved examples of gypsum plasterwork in the pre-Classical period are found in the monumental architecture of ancient Egypt, dating from the 3rd millennium BC; examples include the pyramids of Giza, which contain gypsum mortars, and countless surviving works of frescos and ornament, such as the renowned gypsum bust of Nefertiti. In fact, the gypsum plasters produced in Egypt were in many cases of superior quality to what is commercially available today.

Origins and chemistry

Gypsum is a naturally occurring stone that forms when limestone is exposed to sulfuric acid from volcanic activity. It is easily dissolved, and gypsum deposits meters thick commonly formed through cycles of evaporation in lagoons or dead seas.

Gypsum is a metallic salt of calcium. The most common form of naturally occurring gypsum is calcium sulphate dihydrate, or $CaSO_4{\cdot}2H_2O$. This hydrous gypsum binds water to calcium sulphate molecules in a dry, crystalline state. As we'll see, this imbues gypsum plasters with some unique properties.

Types of gypsum

Unlike clay, gypsum must be baked in preparation for its use as a plaster. Fortunately, this occurs at a relatively low temperature, so it is not an energy-intensive process. Gypsum can be efficiently baked at temperatures as low as 350°F (177°C). At this temperature, gypsum quickly loses 75% of its water content, off-gassing steam. The resulting material has the chemical formula calcium sulphate hemi-hydrate or $CaSO_4{\cdot}\frac{1}{2}H_2O$

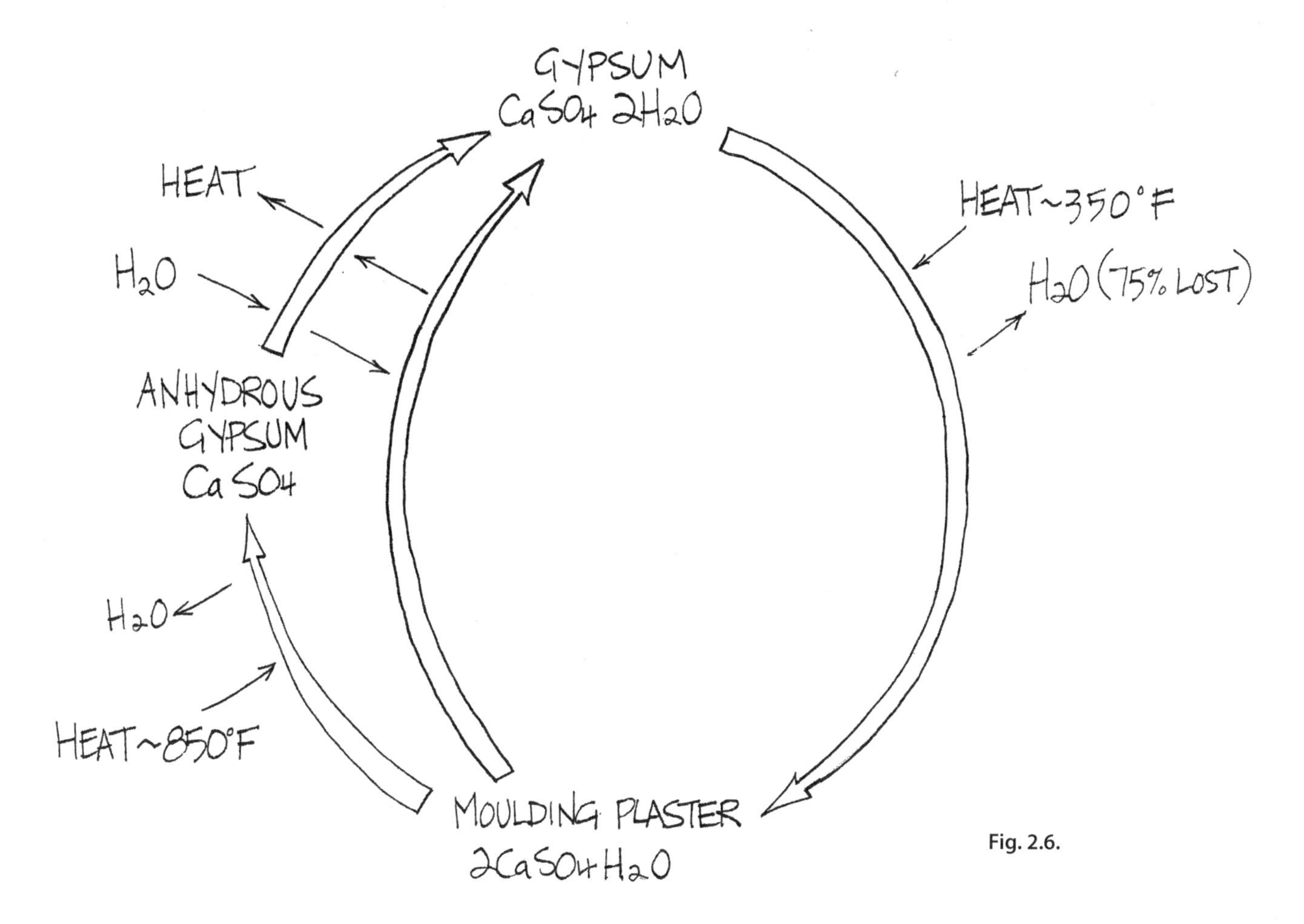

Fig. 2.6.

($2CaSO_4{\cdot}H_2O$). This is commonly known as plaster of Paris, or moulding plaster. *Gauging plaster* is chemically identical to moulding plaster, but it has a coarser grind; it is used with lime plasters to speed up the set.

Anhydrous gypsum can be manufactured by continuing to bake the hydrous form to a temperature of 800–850°F (425°–450°C), producing calcium sulphate or $CaSO_4$. This anhydrous or *dead burnt* gypsum, sometimes with a small addition of alum, is characterized by a slow set and dense crystallization that is useful for floor, exterior, and other specialty applications. Anhydrous gypsum deposits can also occur naturally.

There are several characteristics that are unique to all gypsum plasters. Notable among them is that gypsum is *self-binding*. Aggregates may be added as an inexpensive filler or for decorative effect; however, unlike with clay or lime,

Understanding Drywall and Drywall Compounds

Off-the-shelf drywall mud may contain toxic compounds. It is premixed joint compound that hardens by drying (rather than setting); it isn't really gypsum-based, and should usually be avoided. Drywall itself is less ecological than most gypsum plasters because the gypsum used to manufacture drywall is usually a by-product of pollution control on coal-fired power plants. A great deal of energy is used to process this manufactured gypsum, and there may be heavy metals or other contaminants in it.

Table 2.7: Gypsum Properties and Uses

Molding plaster	$CaSO_4 \cdot \frac{1}{2}H_2O$ (fine grind)	Interior use only.	Very fast set time, primarily used in moldings. Commonly sold as plaster of Paris.
Gauging plaster	$CaSO_4 \cdot \frac{1}{2}H_2O$ (coarse grind)	Interior use only.	Intermediate set time used pure or blended with other binders in plaster.
Dead-burnt gypsum	$CaSO_4$	Interior use, sometimes exterior.	Slow set time blended with other binders in plaster.

they are not necessary for the plaster to hold together. A closely related quality is that gypsum plasters have no shrinkage. As it incorporates most of the added water into its crystalline matrix, it actually expands slightly as it sets; as a result, gypsum plasters have no shrinkage cracking. Plaster of Paris and gypsum cements are fast-setting materials that permit work to be conducted expeditiously. Gypsum plasters have excellent adhesion to most any solid, fibrous, or lath substrate and provide a permeable, breathable coating. Gypsum may be blended into interior lime or clay plasters to reduce shrinkage and provide a rapid initial set.

Although all natural plasters are incombustible, one of gypsum plaster's unique and most cherished qualities is its inherent capacity to actively retard fire. This is due to its hydrous chemistry. Should a fire occur in one room, gypsum will continue to off-gas steam, thus suppressing the temperature on the other side of the wall well below the temperature needed for spontaneous combustion. This arrests the ability of the fire to spread.

Historically, gypsum plasters have been used primarily for interiors. Plaster of Paris produces a plaster that is far too porous and soluble for exteriors. However, there is a long history of exterior stuccoes in Europe based on anhydrous gypsum. As with earthen *renders* (exterior coats of plaster), reasonable precautions need to be taken with overhangs and other flashing details to ensure protection from streaming water, and capillary breaks must be established to prevent water from wicking. Nevertheless, the self-binding nature of the material itself allows a great range of technical and aesthetic freedom. Gypsum stuccoes are very manageable to work as a wall plaster and can be applied up to 1" (25 mm) or more in a single coat. They have a rapid set that permits working in almost any season, so long as there is a brief window of good weather. Furthermore, molding profiles can be run *in situ,* elements and ornamentation can be cast and affixed, and many aggregates can be added just for decorative effect.

Safety and handling

Gypsum is relatively safe to handle. A respirator should be used when mixing, because bagged material may contain small amounts of silica.

Blending Binders

Almost all of the binders are compatible, and they can be blended to take advantage of their combined properties in plasters; however, the ratios can sometimes be very important. One exception is that cement and gypsum are generally incompatible.

Gypsum and lime are blended to make *gauged lime plasters,* which are commonly used as a finish over gypsum plasters, or drywall. Gypsum and clay can also be blended to change the working properties and finish time of earth plasters.

Clay and lime are often blended to make *lime-stabilized* earth plasters, but it's very important to add the right amount of lime — in particular, not too little lime. When a small amount of lime is added to a clay, it causes the clay particles to clump together (called *flocculation*). This causes the clay to act more like silt than clay, and it destroys much of the binding property of the clay. As more lime is added and the pH increases above 11, chemical reactions occur between the lime and the clay, basically creating a hydraulic lime.

Clay, lime, and gypsum can also all be blended in a single plaster — e.g. gypsum can be used as an accelerant in a lime-stabilized earth plaster, but only for interior use because gypsum plasters are highly absorbent and will break down when exposed to moisture.

Sand

Sand provides structure and is very important in many plasters, so the quality of your sand can make the difference between success and failure. Good plaster sand should be sharp, clean, and have a diversity of particle sizes.

Properties and Uses

Because plastering sand should be sharp and angular, some natural sands make poor plaster sand. Beach sand in particular should usually be avoided, because waves have been rounding the sand grains for many years (picture building a structure out of round balls).

Particle size diversity is important to create good structure and to reduce the amount of binder needed. Imagine a bucket filled with softballs — how many golf balls could you add to the bucket without changing the total volume very much? Then how many marbles could you add to that? Ideally, a mix has nearly every grain size so that few large voids are left — this creates a structure that resists movement and requires less binder (because there are few voids to fill). Less binder equates to less shrinkage cracking.

Sand should not contain silt (the particle size between sand and clay), which fills the voids in place of the binder, resulting in weak plasters. Clay can sometimes also cause problems in lime-based plasters. Salt can lead to plaster failure, as well as causing rusting of metal lath or any other metal used in plaster preparation. So, when we talk about sand being clean, we mean free of fine particles and unwanted salt, chemicals, or organic matter.

As a conservative rule, the largest particles in your sand should be no more than half the thickness of your plaster, and preferably no less than one quarter. So, if your plaster coat is ½ inch, your largest aggregate would ideally be somewhere between ⅛ and ¼ inch.

Types of Sand

There are several types of sand that are widely available, so when you call a sand yard or any construction materials supplier, you need only tell them what you want, and it will promptly appear at your jobsite ... maybe. Unfortunately, the definition of sand types allows huge variability (even assuming it is followed correctly), and what you receive on the jobsite will depend on what that supplier carries or what is locally available. Nevertheless, as a rough guideline, the main sand types are *masonry sand, concrete sand, stucco sand* (regionally available), *bagged silica sand,* and *calcite* (white limestone sand).

Masonry sand

Masonry sand has a maximum particle size of 3⁄16 inch (4.75 mm), which should mean that it's a nearly ideal sand for many base coats. Unfortunately, using the ASTM standards for particle size distribution, anywhere from 70–100% of the sand can be less than 3⁄64 inch (1.18 mm). This explains why, when we order masonry sand on a jobsite, sometimes it's a nice mix of coarse and fine sand, perfect for a base coat, and

Fig. 2.7: *Some different sands. Top row (left to right): masonry sand, silica sand, 30–50 mesh calcite; bottom row: quarried calcite, 40–200 mesh calcite, 12–350 mesh calcite.* CREDIT: MICHAEL HENRY

other times it's almost 100% fine sand. Worst of all, masonry sand can be 99% rather fine but have just a few pebbles, making it useless for fine finish coats — unless you take the time to sieve it. With masonry sand, you should look before you buy; however, it is still often the best choice available for a given plaster. Brick sand is commonly used as a synonym for masonry sand; it is often a relatively fine version of masonry sand.

Concrete sand

Concrete sand (technically, "fine aggregate") has a maximum particle size of ⅜ inch (9.5 mm), and 5% should fall between 3⁄16 and ⅜ inch (4.75–9.5 mm). It also must have a variety of mid-size particles, and up to 10% can be finer than 100 mesh. Even though the ⅜-inch maximum is more than we'd like for many plasters, concrete sand may be a good choice for a plaster coat that will be applied over ½-inch thick.

Large aggregate doesn't necessarily interfere with a finish, as it tends to push to the back. However it makes finishing extremely difficult and frustrating if the largest aggregate is the same size as the depth of the finish coat.

Silica sand

Most sand is primarily made of silica; however, fine off-white silica sand (available in 50–100 lb bags) can be useful for some finish plasters. Silica sand is commonly used in earth plasters that are applied in a very thin coats and pigmented (vs. painted). Silica sand is available in various mesh sizes; for finish plasters, you want mesh sizes between about 60 and 100. One hundred mesh or finer would make an exceptionally fine and smooth finish plaster that would be prone to cracking if not applied in very thin coats. Silica sand can be used for sandblasting, so it can sometimes be found in rental centers, as well as being fairly commonly available in masonry supply stores. Finer mesh sizes are often available at pottery supply stores. Silica sand is often quite uniform in particle size; larger mesh sizes are often not that good for plaster, so it's sometimes worth blending different silica sands.

Limestone sands

Limestone-based sands are traditional in many lime finish plasters, especially ones that are highly polished. They are also used in some unpainted earth plasters, where they impart a subtle sparkle. Fine white limestone-based sands are hard to find in bags, but are sometimes used in swimming pools and other types of stucco, so they may be available at masonry or stucco supply stores. You could try going directly to limestone quarries; they may sell you some (though they may have a minimum order size that's more than you want). For finish plasters, quarry sand often needs to be sifted, which means it needs to be dry — that pile of sand stored outside at the quarry looks much less appealing when it's time to sift it. Even bagged sands often need to be sifted for use in finish plasters.

You may encounter the term *marble sand*; often, this is actually calcite sand. The confusion in terminology probably stems from stonemasonry conventions that label some types of good-quality calcite (a sedimentary limestone) as marble (a metamorphosed limestone). So if you buy marble sand or *marble dust,* you may not be buying marble — and that's fine.

Stucco sand

Stucco sand is what you probably want for your plaster, but depending on where you live, the trouble may be actually getting it. It's similar to masonry sand, with a maximum particle size of $^{3}/_{16}$ inch, but has a greater proportion of large particles, and less very fine sand. Stucco sand is ideal for anything but thin finish coats; however, it is far less widely available than masonry or concrete sand.

Sieving Your Own Sand

If you can't buy the sand you need, you can "make" it. It's not unusual to modify sand for use in fine finish plasters by sieving out the larger particles from a commercially available sand. To sieve your own, you need dry sand and an appropriate screen. First, you need to know what mesh size you want (the recipes in this book will specify this, if it's important). Window screen

Fig. 2.8, 2.9: *Sieves of all manners can be used for sieving sand, but it's usually easiest if the sieve is sloped and there's an opening for coarse aggregate to drain.*
CREDIT: TINA THERRIEN, MICHAEL HENRY

is typically about 16 mesh. A smaller mesh size can be purchased in the form of splatter guards from the kitchen section of many stores (maybe around 30 mesh), otherwise, you will likely be shopping online at a specialty supplier (see Appendix 2, Resources).

It's worth taking the time to make a good setup for sieving sand — this may be as simple as cutting the bottom out of a bucket and gluing mesh in (construction adhesive is good for this), but it usually involves making an angled wood frame. For anything but small, occasional sieving jobs, it's worth investing time in a good setup. Build the frame to fit on a slope over a Rubbermaid, or whatever bin you find most convenient. Leave the lower end open for coarse aggregate to exit and hook a bag over the end so that it bags itself. Try out some different angles (you can start with 45) until you find the one that works best for you.

Most sand is primarily made up of silica, so avoid breathing the fine dust — always wear a respirator to sieve sand.

Fiber

Fiber is used in many, but not all plasters. The role of fiber is to increase tensile strength and ductility — meaning it helps plaster resist stretching, compression, and deformation. It can also play an important role in reducing shrinkage cracking.

Types of Fiber

Natural fibers used in plasters include straw, hemp, animal hair, coconut, sisal, flax, cattails, wild rice hulls ... and many other fibers. Straw is the most commonly used fiber, and it can be used in a wide variety of plasters. It can be cut at different lengths to use in base coats or finish coats. It's not as commonly used in finish plasters unless they will be applied thick or the look is desired — fine straw in a finish can be very beautiful.

Fiber has several function in earth plasters:

- It allows for increased depth of application. Plasters that are meant to be applied very thick

Fig. 2.10: *A small selection of the many possible fibers that have been used in plasters: 1) coarse chopped straw 2) mulched straw processed with a wood chipper 3) hemp sliver 4) textile flax 5) wild rice hulls 6) polypropylene fiber (synthetic).* Credit: Michael Henry

will have a lot of fiber, whereas many thinner finish plasters have no fiber.

- It compensates for substrates that are prone to movement, and also helps the plaster bridge a change in substrate without cracking along the transition.
- It allows for changes in depth, filling hollows in the wall, and building raised sculptures or other structural elements on the wall.
- The amount of fiber in a plaster usually ranges from 0–30%, but it can be more. Generally, the thicker a plaster will be applied, the coarser the fiber that will be used in it.

Straw

The fiber of choice for most earth plaster base coats, and some finish coats, is chopped straw. Where the base coat will be very thick (to level uneven walls or fill voids, for example), the straw is usually chopped to 2–6 inches in length. However, coarse straw reduces workability of the plaster, so for thinner base coats, or where a relatively fine finish is desired, straw can be chopped to as little as 1–2 inches in length. Finely chopped straw will contribute less to the strength of the plaster, and it may be more prone to cracking when applied very thick, but adding a larger amount of fine straw to a recipe can often substitute for a smaller quantity of long straw, with better workability. A *mix* of different lengths of straw often produces the best results.

In straw bale homes, floor sweepings from window trimming or weed whacking the walls can be a good source of fine to medium-length chopped straw for plaster. Bag them in contractor-grade garbage bags and save them for later. Or, to make coarse chopped straw, simply carve up a bale with a chainsaw: with at least one string still tied, cut the bale down the entire length, carving off 2–4 inch strips. You may be tempted to cut with the top of the bar to avoid pulling too much straw into the mechanism. If you do this, just keep yourself, and everyone else, well clear of the saw — in case it kicks back. When you use the top of the bar, many of the chainsaw's safety features can't help you. Proper chainsaw safety gear should be worn.

Finer straw usually needs to be screened. Using a wood chipper is often the most efficient way to process a lot of fine straw. Otherwise, you can use a leaf mulcher, or improvise a mulcher using a weed whacker and a tub. Cut the bottom out of the tub (leave a rim), and duct tape metal screening or fencing with the desired opening size into the hole. Half-inch to one-inch fencing will screen a nice fine-chopped straw. Cover the top of the tub with plastic, with a slit to feed in straw and accommodate the shaft of the trimmer. You'll find that the trimmer mulches the straw and at the same time pushes it through the screen much more efficiently than you could do it by hand. Set the mulching tub in another tub with a spacer to leave a cavity where the straw can land. A half hour invested making this contraption can save hours of labor on a job. Use a trimmer that has its motor on the back — if the motor is inside the mulching container, it will burn out right away.

Tom Rijven uses a lawnmower to mulch straw. This involves running a mower (without a side-chute, or bag) over the straw a few times, in a long wood frame with a plywood base.

Straw contains significant amounts of silica (wheat straw is 3–5% silica, rice straw has more), so a respirator should always be worn when working with dry straw that could create dust.

Hemp

Hemp fibers vary a lot in their properties depending on what part of the plant they come from and how they are processed. Hemp hurd, or shiv (the core of the stem), is a coarse material that is typically used in sculpting and for building structure, but it is otherwise not very useful in plasters. Hemp *sliver* is a fine fiber that can be used in finish coats or as an addition to thicker coats. It may be subtly visible in finer finish coats. These fibers tend to collect on the paddles of most types of mixers unless they are cut fairly short, usually ½ inch. This type of fiber may be available from paper-making suppliers or other online retailers.

Flax

Flax (*Linum usitatissimum*) is the plant from which linen is made, and its seed is the source of linseed oil. Flax is an underrated plant that can play a very valuable role in natural building. Flax can be purchased in a variety of qualities and types from paper-making supply shops. The most useful for finish plasters are *textile flax* or flax *noils* (the leftovers from combing machines), which can be used as very fine fiber in finish plasters.

Cattail fluff

Cattail fluff can be used as a very fine fiber in finish plasters. It should be collected when it is loose and dry on the stalks, and a little of it can go a surprisingly long way. It is a very short fiber, so is not suitable for plaster that will be built to any significant depth. However, because it essentially vanishes into a plaster, it allows for fine finishes.

While some of the fibers in this section don't appear in recipes in this book (cattail, coconut, and flax), we have successfully used cattail fluff and flax in recipes in the past. Feel free to experiment with substitutions of fibers in your mix, to make sure they have the desired effect. Locally available fibers are cost effective and easily obtained.

Synthetic fiber

Synthetic fibers used in plaster are primarily either fiberglass or polypropylene. Synthetic fibers are commonly used in lime and cement-lime plasters, but they can be used in earth plaster, too. Polypropylene is the most likely choice because it works amazingly well in small quantities and is readily available from most masonry supply stores. However, while it doesn't show at all in a troweled finish, it typically shows when a sponge finish is used (because it lifts out and leaves little hairs on the plaster surface). This is a problem, especially in unpainted finish plasters. Natural fibers such as fine hemp, flax, or cattail fluff are comparable to synthetic fibers, but using them means you avoid putting plastic into your lovely earth plaster, and the surface can be sponged without having the fiber lift out of the plaster. All things being equal, go natural — but in a pinch, or for certain applications, synthetic works well with very little material.

Pigments

Pigments are mixed into many fine finish plasters, as well as natural paints and alis (clay paint). Many natural pigments are simply clays that have high proportions of iron oxide or other minerals that lend their characteristic color. Pigments vary in their toxicity, UV-stability, and lime (alkali) stability.

Types of Pigment

Earth pigments

Earth pigments are rocks, minerals, or clays that are dug directly from the earth. If they are not already a fine powder, they are ground into one. Ochre is a very well-known earth pigment with a color that comes mostly from hydrated iron oxide (limonite). Umber and sienna derive their color from hydrated iron oxide and manganese oxide.

When umber and sienna are heated, the limonite is dehydrated, converting some of it to a much redder hematite. The resulting pigments, which are darker and more reddish-brown than the raw forms, are known as burnt umber and burnt sienna.

In many cases, earth pigments have been mined from the same areas for centuries and can have historical and cultural significance.

Synthetic pigments

Synthetic pigments may emulate pigments that were traditionally derived from natural sources, or they may be novel inventions. Many of the pigments of interest to plasterers are inorganic pigments; however, some organic pigments can also be used in plaster and can offer extremely stable, vibrant colors.

Mars was originally a brand name in the late 18th century, but the term is now often used for any synthetic iron oxide pigment. Mars yellow is a synthetic yellow ochre; when heated, it transforms to a wide variety of other colors including Mars orange, violet, and black. Even though they are synthetic pigments, Mars colors are usually grouped with earth pigments, to which they are closely related.

Many of the other synthetic inorganic pigments are oxides of other metals. Chromium oxide green is a common pigment that is quite beautiful. Chromium oxides have a bad reputation because of chromium trioxide, a toxic form of chromium used in electroplating. The form of chromium oxide used as a pigment is considered safe and nontoxic.

Naples yellow is a lead-based pigment and is therefore highly toxic and should be avoided. However, there are replacement pigments that may use the name Naples yellow (e.g. *Naples yellow deep*) but contain no lead — these have the pigment number Pbr24. This pigment group includes some beautiful yellow and yellow-orange colors.

Pigment Stability

Some pigments break down quickly in the presence of lime (they are not *lime-fast*). For example Prussian blue, a deep blue pigment, turns almost completely white within minutes of being mixed with lime. Artists who work with

Table 2.8: Pigment Characteristics

Some Common Pigments and Pigment Groups	Lime Stable?	UV Stable?	Toxicity	Notes
Earthtones including ochre, umber, sienna, iron oxides Py42,43 Pr101,102 Pbr6,7,9 etc.	Yes	Yes	Low	May contain silica, otherwise nontoxic. Affordable, widely available.
Cobalt colors Pb26,28,36,50	Yes	Yes	Low	Expensive.
Chromium oxide green Pg17	Yes	Yes	Low	Safe and stable oxidation state of chromium.
Pbr24 (range of colors)	Yes	Yes	Low	Chrome antimony titanate. Naples yellow alternative.
Mayan pigments	Yes	Yes	Low	Organic dyes chemically bonded to clays.
Hansa yellow Py3,74	Yes	Yes	Low	Synthetic organic pigment.
Indanthrone blue Pb60	Yes	Yes	Low	Synthetic organic pigment, very stable but expensive.
Naphthol reds Pr170	Yes	Variable	Low	Synthetic organic pigment.
Pigments with Some Toxicity				
Ultramarine blue Pb29	Variable	Yes	Low-moderate	Some ultramarine blues are marketed as limefast.
Cadmium colors	Yes	No	Low-moderate	Reports of toxicity vary from nontoxic to carcinogenic.
Titanium dioxide	Yes	Yes	Low-moderate	Possible carcinogen if inhaled. Nano-sized particles may be more toxic than larger ones.
Phthalo colors	Yes	Yes	Low-moderate	May damage kidney and liver at moderate doses, but this is not clear. MSDS sheets vary in reporting toxicity.

fresco know which pigments they can use with lime, and they often post good information on the internet. To test for yourself, mix the pigment with lime putty and leave it mixed wet in a jar for a few weeks to see if it fades.

Some pigments fade in sunlight (they are not UV stable). For instance, cadmium colors are stable in lime, but not UV. Mural painters have lists of colors that are very stable outdoors.

Maximum Quantities

The maximum pigment that should be used in plasters is often given as 5–10% of the binder by weight. Just to confuse things, the same ratio by volume is often cited as a maximum; in reality, it will vary a little depending on the type of pigment. Regardless, if you're aiming for the higher end of the range, evaluate your plaster tests — weak or dusty plaster may have too much pigment. Three to five percent often gives a nice color.

Toxicity

Don't assume pigments are safe. Make sure of it by checking the safety data sheet (SDS or MSDS) for every pigment you work with. Even pigments that are considered to have low toxicity should never be inhaled, as they may contain small amounts of silica or other products that can be dangerous as a very fine powder.

Mixing Pigments

Pigments need to be properly mixed in order to avoid streaking. Some pigments (e.g. ochres) can be mixed into the dry mixture before adding water. However the best way to mix any pigment is as follows:

- Blend the pigment with a small amount of water, roughly an equal volume to the pigment.
- Stir this mixture well; it should be a workable paste — if it's too thick, add water.
- After stirring well, slowly stir in more water until it's the consistency of thin paint.
- Add the pigment to the mix.

To mix the pigment evenly into the plaster:

- Mix half the dry ingredients into the water.
- Add the pigment slurry and mix well.
- Add the remaining dry ingredients and mix again.

Blending Pigments: Color Theory

An understanding of color theory is beyond the scope of this book (a good reference is *The Keys to Color* by Dean Sickler), but if you work with pigmented plasters, you'll have to confront color mixing. A useful concept is *modifying color by adding complementary colors.* Complementary colors are opposite each other on the color wheel, and when they are combined in a paint or a plaster they cancel each other out — so it's a subtractive approach to using color rather than the additive one that one normally thinks of.

For example, if you want to shift a blue-green toward the green, rather than adding green, you could add orange, the complement of blue, to cancel it out. However, if you wanted to tip it more toward blue, you would add red, which would cancel some of the green. This trick is especially useful when working with natural pigments, where the color palette is limited (maybe the hue of green you want to add doesn't exist). Of course, sometimes the best way to get more green is to add green!

Chapter 3

Planning and Preparation

How to Choose a Plaster

THE BEST PLASTER FOR A BUILDING will depend on many things, including the type of construction, how it is prepped, the local climate, and the design of the building. There may be one obvious choice — or a number of equally good choices. You should talk to a local builder and/or owner-builders about their experiences. Remember that local vernaculars usually exist for a reason — if you're going to deviate from them, do your homework.

Generally, plasters fall on a spectrum of *soft* to *hard*. As hardness and strength increase, permeability and flexibility decrease (as a general rule). Therefore there is no "best" plaster; there is only the right compromise for your building. Let's take a closer look at the attributes of plaster.

Permeability, Absorption, and Adsorption

Vapor permeability describes the ability of a plaster to let moisture pass through. Higher permeability is nearly always better because water will eventually find its way into a wall somehow (through cracks, flashing errors, air leaks, condensation, etc.), and the water needs to be able to find its way back out again. Inviting small amounts of water in through earth or lime plasters may seem like asking for trouble, but natural wall systems can usually uptake quite a bit of moisture without reaching a threshold where mold can grow — as long as they are able to dry out again. There are surprisingly few studies of vapor permeability of natural plasters.

Absorption is a familiar concept, but *ad*sorption is not: it is the physical binding of particles of gas, liquid, or solid onto the surface of molecules. Some plasters, especially earth plasters, can adsorb large amounts of water vapor without becoming damp or allowing any mold growth. Earth plasters regulate humidity by adsorbing water vapor when the air is above 50% relative humidity, and releasing it below that threshold. Earth plasters also *ab*sorb water readily — up to a threshold. These combined properties allow earth plaster to protect other natural materials such as wood or straw from moisture damage, by wicking moisture away.

Strength and Flexibility

There are different ways to measure the strength of plasters, so it's hard to make broad generalizations. In most ways, strength decreases as you move from cement to lime to earth. However, earth plasters have pretty high compressive strength, which is important if you are planning a straw-bale load-bearing building, or if your plaster may carry extra load.

The softer plasters also tend to be more flexible, and that is usually a good thing in natural building. If you're plastering over natural materials such as straw or earth, you want your plaster to have a flexibility similar to the substrate, so that it can move with it. Also, the more flexible plasters deal better with expansion and contraction, which are the inevitable result of drying and temperature changes. All in all, this means fewer cracks, which means less maintenance on your finished home, and fewer potential entry points for water.

Finding the Right Compromise

So there are trade-offs, but on the whole, we think the balance tips in favor of earth; it's more flexible, repairable, and — most importantly — it's permeable. Why not just plaster everything with earth then? In dry climates, you might be able to. In wet climates, exposed earth plaster will wash off a wall relatively quickly. Driving rain is the enemy of earth plasters, and they need to be protected. One way to do this is by incorporating extreme overhangs into your design, such as a porch roof that runs all the way around the house.

Another option is to avoid exposing earth plasters to the elements by using lime as a finish plaster on the outside of buildings. Lime has a good balance of strength, permeability, and flexibility for an exterior plaster. Lime is still porous enough that it needs to be protected from rain by an appropriate finish, but it resists being worn away by the weather. Some of the disadvantages of lime come from the difficulty of applying it. Lime is finicky. If it is applied too thick, or in hot weather, or if it is allowed to dry too quickly, it can fail. This may mean putting on extra leveling coats to build up uneven walls, and it often means postponing plastering until cooler weather.

One way around some of the challenges of working with lime is to use an earth base coat and a lime finish coat. Unfortunately, where this has been tried, the plaster has often delaminated, creating a hollow space within the plaster where moisture can accumulate — and freeze. In time, parts of the finish coat are likely to fall off. The solution is to stabilize the earth with lime so that the plaster coats behave in a similar fashion. The amount of lime added is important: too little, and the plaster is likely to fail.

Table 3.1: A Comparative Table of Binders

Binder	Permeability (US Perm/Inch)	Weathering Resistance	Strength	Embodied Energy
Clay	Excellent (18)	Poor	Weak	Low
Gypsum	Excellent (18)	Very poor	Weak	Low-medium
Lime	Good (14)	Good	Strong	Medium
Cement-Lime	Poor (7–10)	Very good	Strong, but brittle	High
Cement	Very Poor (1)	Very good	Strong, but brittle	Very high

Soft and Hard Plasters: A Few Rules

When we talk about a soft plaster, we mean one that is more flexible and permeable to moisture; the softest plaster would be an earth plaster. Hard plasters, which are more rigid and vapor impermeable, include hydraulic lime and cement. Let's review a few rules for matching materials, but each rule will break the one before.

Rule #1: Like bonds to like.
Generally, we want to match materials, so if you're plastering over an earth building, the ideal plaster is earth; over concrete, a cement stucco is most appropriate. Straw bales are a fairly soft substrate, so earth or maybe lime would be appropriate. The like-to-like rule also applies to plaster coats: ideally, we want all our coats to be the same material.

Rule #2: Soft over hard, not hard over soft.
When you have to break Rule #1, don't put harder coats over softer coats. The reason for this is that the upper coat will tend to trap water, which will, over time, damage the coat below. Also, the coat below is more flexible and will move with the substrate more readily than the finish coat will — this could result in delamination between the coats, and risk the upper coat falling off.

Rule #3: Increase hardness in small increments.
Rule #2 is a problem because we want the most weather-resistant coat on the outside, so there is a long history of breaking this rule. However, it must be done with caution — if/when you break Rule #2 by putting hard over soft, it's essential to keep the materials as similar in properties as possible. One way to bridge the difference is to put on several coats, with each increasing in

hardness. But don't go overboard — for example, no matter how many layers one uses, cement can't go over earth. But lime can, especially hydrated lime. The intermediate coat in this case would be lime-stabilized earth.

Mechanical key

Physical bond between coats is always a good idea; when using dissimilar materials, it's generally essential. For very thin coats, a rough texture might be enough, but for thicker coats, a deep scratch pattern will help the upper coat *key* into the base coat.

Some Common Exterior Plaster Systems

1. *Hydrated lime finish coat over lime-stabilized earth base coat.*
   This is a good system for maximizing permeability with fairly good weather resistance. Should be painted with silicate or other vapor-permeable paint. Plastering season: Base coat in summer, finish in early autumn (or late spring).
2. *Hydrated lime base and finish coats.*
   A vapor-permeable plaster that is usually built up in three coats. Should be painted with silicate or other vapor-permeable paint. Plastering season: early autumn (or spring).
3. *Hydraulic lime base and finish coats.*
   A moderately vapor-permeable plaster with good weather resistance and less absorbency than hydrated lime. Primary impediments are added materials cost, and added embodied energy of transport. Plastering season: spring, cool periods of summer, autumn.
4. *Earth plaster base and finish coats.*
   Earth plaster can be used as an interior finish in any climate. As an exterior finish, in warm, dry climates it can be used as a sacrificial layer with the intention of simply re-plastering it when it wears away. It can also be used in any climate with silicate paint for weather protection — if it is fully sheltered, by a porch roof etc. Plastering season: summer.
5. *Base coat of earth or lime covered with rain-screen or siding.*
   On exposed sites (such as hilltops), siding will usually offer the best weather protection; however, many natural wall systems need a base coat of plaster to seal the wall to reduce air leakage, to protect against pests, etc.

Natural Plasters over Conventional Construction

Stucco is a popular alternative to siding in many parts of North America. Traditional stuccos were cement-lime or lime, which means they had a relatively high vapor permeability. These stuccos are still sometimes used, but they have been largely replaced by modern stuccos, which are commonly acrylic-based and have very low vapor permeability. Stuccos with high lime content, and without synthetic additives, can help protect the building by allowing the sheathing below to dry to the outside more easily. Hydraulic and hydrated lime plasters have been used over conventional construction with good results.

Modern construction techniques such as sheathing houses with oriented strand board (OSB), installing lots of insulation in the walls, and the use of interior vapor barriers have led to an epidemic of rot issues in many homes, especially those with stucco cladding. When moisture gets trapped beneath any stucco system (even a partially vapor-permeable one), it can rot the wall sheathing, particularly if that sheathing is OSB. Fortunately there's a simple solution: create a small air space and drainage plane behind the plaster. A variety of products have been invented to do this; they are described in more detail later in this chapter in "Water-resistive barriers and drainage planes." Drainage planes are recommended over sheathing for all plaster types.

Designing for Your Plaster

Siting of Building

Where you put your building will affect everything. The best building sites have low exposure to wind to reduce the likelihood of driving rain. You may want to site your building on top of an exposed hill for the view, or to keep bugs away, but be aware that this will limit your design options. Orienting your building correctly can help mitigate exposure, as can planting trees on windward sides to create a windbreak. Pay attention to the direction of prevailing winds as well as storm winds.

Good building sites must also have good drainage to reduce the risk of flooding, or at least there must be the potential to plan and install good drainage, and to grade the slope properly away from the home.

Building Design

Good building design is essential for the success of any natural plaster. The most important design detail is probably the roof overhang. Overhangs will be specific to the building shape and the exposure of the site, but as a rule, one-story buildings should have a minimum 16-inch overhang, and two-story buildings at least 24 inches. The weak point in many natural buildings is gable ends, because a roof overhang there is much less effective. Avoid a large gable end on a windward side of the house. Gable ends in general should include some kind of skirt roof, and you may want to consider siding the upper part if it's large or particularly exposed.

There is a tendency to gravitate toward harder plasters on exposed sites to prevent erosion risk, but be aware that the more durable plasters

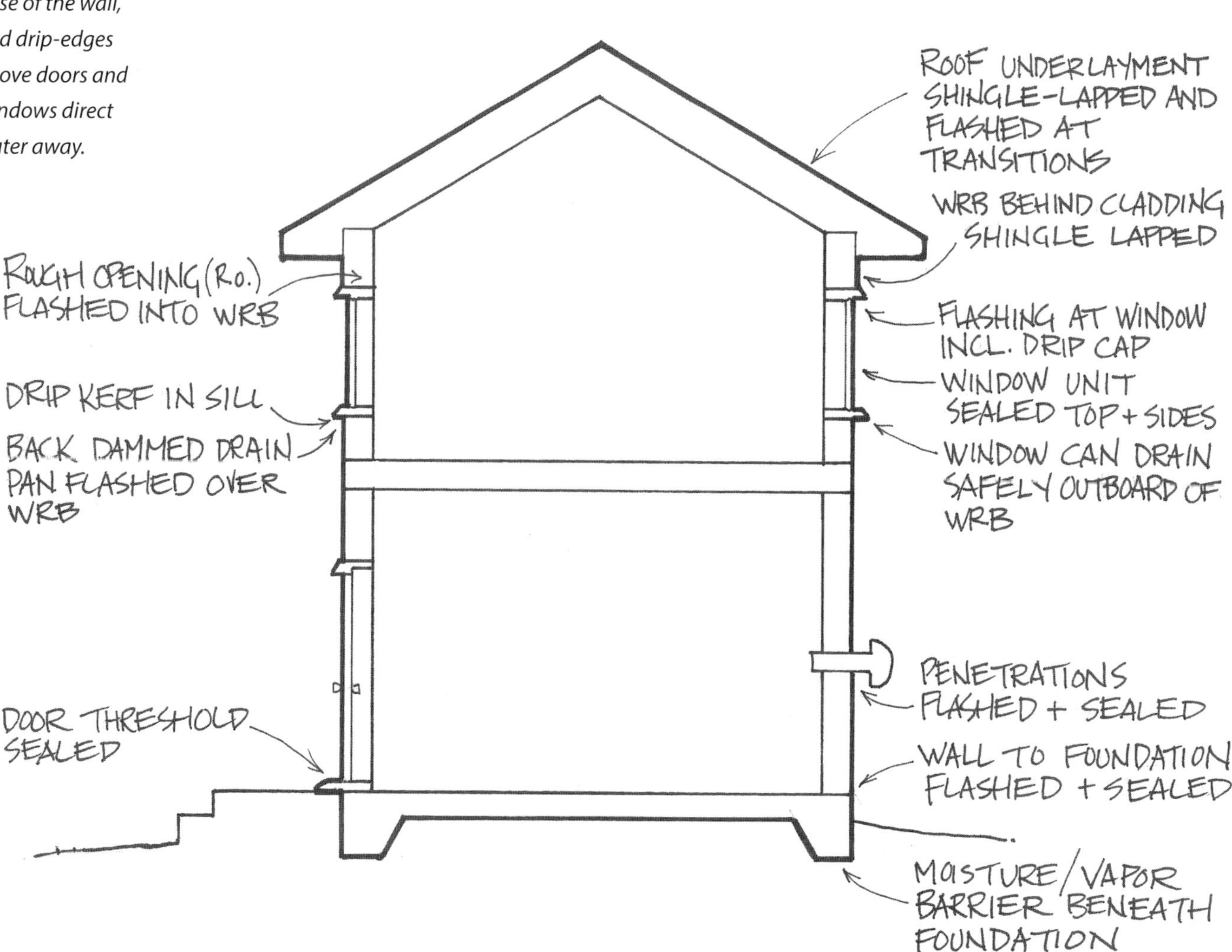

Fig. 3.1: *A good hat and boots are necessary for any building. Flashing is installed at the base of the wall, and drip-edges above doors and windows direct water away.*

are more likely to trap moisture in the wall; if you're worried about your plaster eroding under driving rain, you may have a design problem. One often overlooked alternative that can eliminate most external moisture issues — even on exposed sites — is to use siding or a rainscreen over natural walls. On some wall systems, such as straw bale or slip straw, a base coat of plaster is still needed to seal the wall before siding is added. Venting and drainage should be carefully considered in the planning and installation of rainscreens. *The Natural Building Companion,* by Jacob Deva Racusin and Ace McArleton, offers a good discussion of rainscreens and other design details.

Splashback and flooding can cause damage at the base of walls. Install gutters to direct water into your drainage system; a pony wall or stem wall that raises your natural plaster 18 to 24 inches above the finished grade is a good idea. Proper flashing details at the base of the plastered wall, and around all openings, are also essential to protect your plaster and your wall. A bead of caulking is a final protection at all plaster edges, particularly the base of the wall.

Two- and Three-Coat Plaster Systems

Preparation of natural substrates

- Fill holes: deep holes may be filled with straw or slip straw; surface holes may be better filled using a cob mix.
- Overlap all transitions, including framing in the wall, with either fiber mesh or lath (described below).
- Ensure that air barriers, water barriers, flashing, and plaster stops are in place.
- Very smooth areas may need to be mechanically roughened or covered with an adhesion coat or lath.

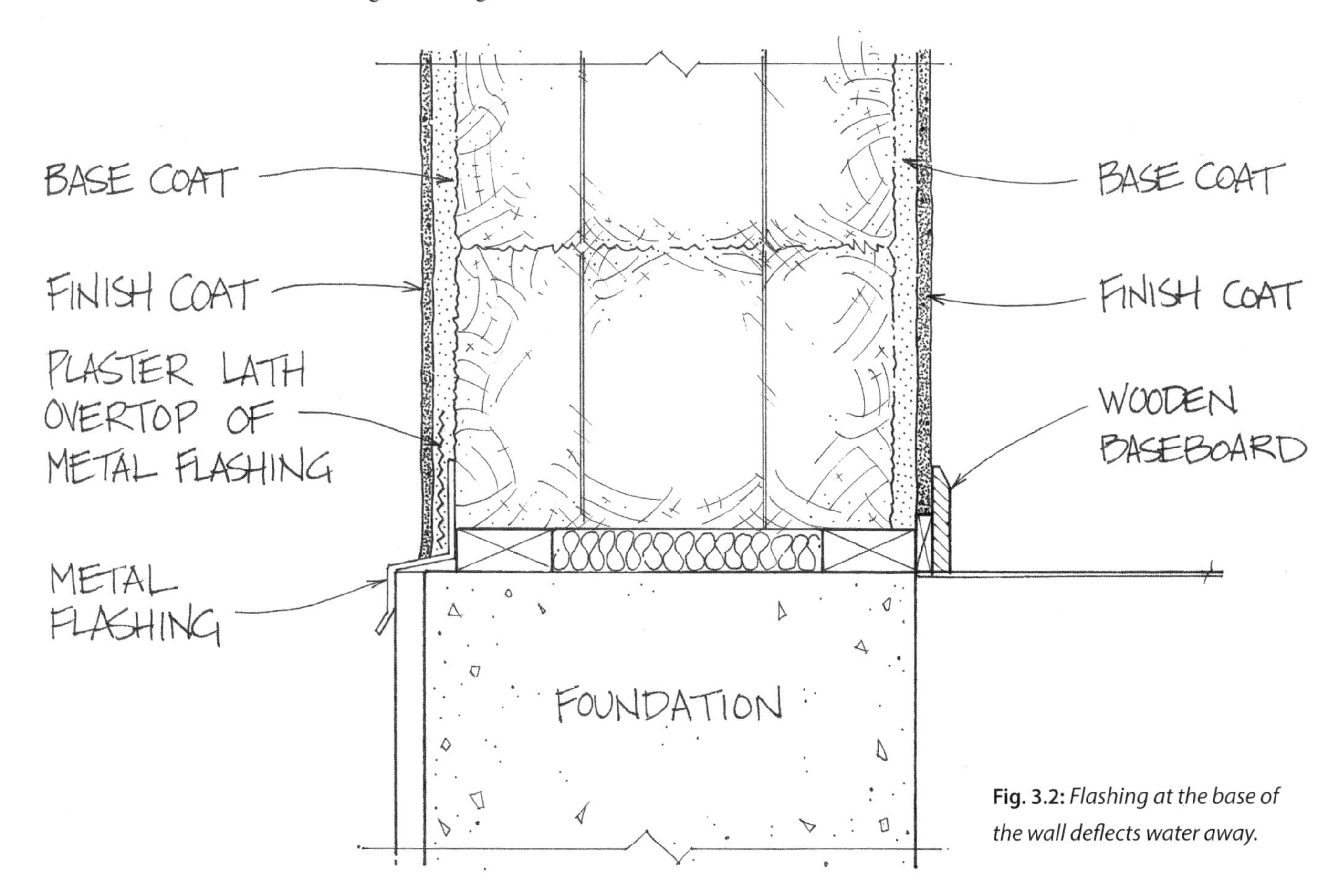

Fig. 3.2: *Flashing at the base of the wall deflects water away.*

Review of substrates: hempcrete, cob, etc.

Cob should be shaped and trimmed as much as possible while still damp. Avoid leaving very smooth areas; roughen them, if necessary, before plastering. Pre-fill any deep holes, and remember to mist cob walls well before plastering. Suitable plasters include earth or lime (hydrated or feebly hydraulic). Cement must never be used over cob. A lime-stabilized earth base coat should be used as a transition between cob and lime finish plasters.

Hempcrete and slip straw (light clay straw) are ideal substrates for plastering. If done well, they are solid substrates with plenty of texture for plaster to key into. Suitable plasters include earth or lime.

Straw bale walls vary tremendously in their quality as a plaster substrate. Good bales, with proper detailing of framing etc., will be satisfying to plaster over. If bales are poor, it may be worth considering the use of plastic mesh. Any plaster may be used over straw bales, but cement-based plasters are not recommended.

Stone is a traditional building material, but it can be quite variable in its properties. Very soft, permeable stones will accept plaster directly without a bonding coat. Soft plasters (earth or hydrated lime) should be matched to soft stone. Hard, less permeable stone may require some kind of bonding coat, and a hydraulic lime is better suited for plastering over it.

Brick is similar to stone — older, fired clay brick is similar in properties to soft stone, while many modern bricks have properties more akin to hard impermeable stone.

Use of plastic mesh over straw bales

In many cases, covering straw bale walls with a plastic mesh is not strictly necessary; however, it can improve walls when bales are poor, and it speeds up plastering, especially when using lime and cement-lime plasters. It can work well with high-sand earth plasters applied fairly wet, but high-straw-content earth plasters may not adhere well because the mesh prevents the plaster from bonding with the substrate. When using mesh over bale walls, it's essential it be well stitched, without voids between the mesh and the straw.

Dealing with framing and transitions

There are many approaches for dealing with framing and other substrate changes in your wall, but here are a few options that we would recommend:

1. Staple metal, plastic, or fiberglass lath to cover wood framing, and extend 2–3 inches beyond the transition edge. This is the best option for lime plasters.
2. Use glue-slip to bond burlap to the framing, and cover transitions.
3. Embed burlap or other fiber mat into the base coat extending several inches beyond

Table 3.2: Stapled Lath vs. Burlap

	Stapled Lath	Burlap (Embedded and/or Glue-Slip)
Advantages	Fast, easy, one-step solution that creates a good bond to structure.	No need for metallic fasteners; easy to source materials; low embodied energy
Disadvantages	With earth-based plasters, a plastic or fiberglass lath must be used, which may be hard to source. Higher embodied energy.	Tends to be slower; may add an extra step during plastering.
Primary uses	For lime, cement, lime-stabilized earth, and earth (nonmetallic lath required for clay-based plasters).	For earth plasters.

the transition (may be used in combination with option 2).

Using lath to bridge transitions

For any two- or three-coat plaster, it's advisable to use a lath that is designed to hold the weight of a ⅞-inch cement stucco. For lime and cement-based plasters, metal lath may be used, but for any earth-based plaster it is very important to use a nonmetallic lath, which may be either plastic or fiberglass. Brand name options include Ultra-Lath, PermaLath, or SpiderLath; these usually need to be special ordered through masonry supply or stucco supply stores, or directly on the internet. Nonmetallic lath is easier and more pleasant to work with, but it tends to be harder to obtain and often costs a little more than metal lath. Hopefully, this will change once it becomes more common. To span transitions with lath:

- Use a hammer tacker to attach strips of house wrap over wood framing (no overlap to adjacent materials).
- Cut strips of lath wide enough to overhang a minimum of two inches on either side of framing and attach with air stapler.
- Attach edges of lath to straw where needed, using landscape staples or other pins.

Using burlap and other natural meshes

Burlap can be soaked with glue-slip and adhered directly to the wood framing (staples are optional). Another solution is to embed burlap in the base-coat plaster to span changes in substrate. This won't work for spanning larger areas of wood, but it can be used in combination with glue-slipped burlap for these areas.

Glue-slip adhesion coats

Make the clay slip a milkshake-type consistency and then slowly paddle in the PVA wood glue

Fig. 3.3: *A straw-bale wall prepped with fiberglass lath. Air sealing tapes surround penetrations, and vapor retarder (covered with mesh) extends several inches over bales at the top and bottom of the wall, to act as an air barrier.*
CREDIT: DEIRDRE MCGAHERN

(about 1–2% by volume) until it begins to look more viscous, with a latex paint consistency. If you go too far, it becomes more of a pudding consistency, which is more difficult to work with a brush or when dipping the burlap. (Thanks to Liz Johndrow for sharing this technique.) Paint the mix onto the wood framing then apply burlap dipped in the same mix (overhang to bridge transitions, unless embedding burlap in plaster coats). Staples are optional for small surface areas, but recommended for larger surfaces or key areas, such as tops of walls or window tops; stapling can be done after slipped burlap is dry.

Plastering over Wood Frame Construction

Sheathing

Plywood and OSB sheathing should have a ⅛-inch gap between sheets to accommodate expansion. This is generally specified in the building code, but not always followed; it is a particular concern to plasterers because buckling at joints may cause plaster to crack.

Water-resistive barriers and drainage planes

Some kind of water-resistive barrier is needed between the plaster and conventional wall sheathing such as plywood or oriented strand board. Standard Tyvek housewrap has a reputation for degrading and losing its waterproofing when exposed to some plasters (and even some types of wood), while TYPAR seems to perform better. Tyvek also makes a stuccowrap product that is more durable. Asphalt felt (double layered) is often written into building codes for use under stucco, and it resists water, is somewhat vapor permeable (5 perms when dry, but 60 perms when wet), and is likely more durable than many housewraps. In some cases, this may be adequate, but a drainage plane is strongly recommended, and it may be a code requirement.

A drainage plane is simply a gap between the plaster and the sheathing that allows any water that finds its way behind plaster to escape. The easiest way to achieve this is to install *drainage mats* (a mat of coarse plastic fibers) or some other material to create an air gap between plaster and underlying materials. Various commercial products are available: MTI Rainscreen, Korax Rainscreen Panel, GreenGuard Drainage Mat, WaterWay Rainscreen Drainage Mats, Korax Stucco Rainscreen Panel, Slicker MAX rainscreen, the Mortairvent Mortar Deflection and Ventilation System, Delta-Dry, are just some of the products you might run across. We don't recommend any particular system; do your research and use one that you can obtain locally. Make sure that drainage is provided at the bottom of the wall for moisture to escape. This may be achieved using an angled weeping screed, J-trim/casing beads with weeping holes, or other means of creating a gap for drainage.

Installing metal lath (or alternative)

Natural plasterers often find themselves installing significant areas of lath, most commonly on gable ends. Too often, it is improperly installed. Here are a few rules of lath installation:

- Ensure that a water-resistive barrier, drainage plane, and the required flashing are in place.
- Use lath that is rated for the correct depth of plaster.
- Use exterior-grade lath where needed.
- Use staples or roofing nails that are a minimum of 1 inch long.
- Fasten to structural members, not sheathing.
- Space fasteners every 6 inches on 16-inch centers, or every 4 inches on 24-inch centers.
- Overlap horizontal and vertical edges as specified by your building code (commonly 1–2 inches). Larger overlaps may be detrimental.
- Stagger end joints of lath.
- Wrap corners by at least 6 inches (either wrap each sheet or add vertical strips on corners).

Installing wood lath

Wood lath is a traditional substrate for plaster, and it still may make a lot of sense for interior frame walls in natural homes or as a base for plaster in homes where chemical sensitivities may preclude the use of many drywall materials. Freshly cut green wood isn't ideal to use as lath; wood that is still "slightly green" is acceptable, however. For small areas, we generally make our own lath by running 2× 4 lumber through the table saw to make strips, but commercially produced lath can be ordered in many areas.

Here are some tips for installing wood lath:

- Use good-quality lath that is dry, not green. Traditional lath strips were commonly 1.5 inches × ⅜ inch × 48 inch long, and they were made of a variety of softwoods. For many projects, ¼-inch lath is also acceptable.

- Wet the lath frequently for a day or more before, and the day of plastering. This is commonly done by misting the installed lath, but some plasterers immerse the lath in water for a full day before installation, then keep it misted until plastering. On the other hand, heritage plasterer Ben Scott simply suggests that we "hydrate the lath slightly with water before applying the plaster." Though there are a few ways to do it, the idea is that the lath needs to be dampened to prevent it from absorbing water from the plaster and expanding while the plaster is shrinking. If it isn't adequately wetted, there will be more shrinkage cracks in the plaster, and it could affect the bond between lath and plaster.
- Space strips of lath with a ¼-inch gap on walls and a ⅜-inch gap on ceilings. Keep gaps even and consistent. Much larger gaps (up to an inch or even more) can be used when straw-rich earth plaster base coats will be applied — but do ample testing to be sure!
- The ends of the lath should always land on studs. Leave a ⅛–¼ inch vertical space between the butt-ends of lath.
- Stagger joints periodically. Apply only 5–8 courses of lath (up to 14 inches) before moving the butt joint to a different stud.
- When lathing over areas of solid framing, use pieces of lath as furring strips to create a gap where the plaster can key.
- Keep lath damp until plastering using a sprayer.
- Use a plaster mix that has enough fine-medium textured fiber to help with keying, and apply it fairly stiff.

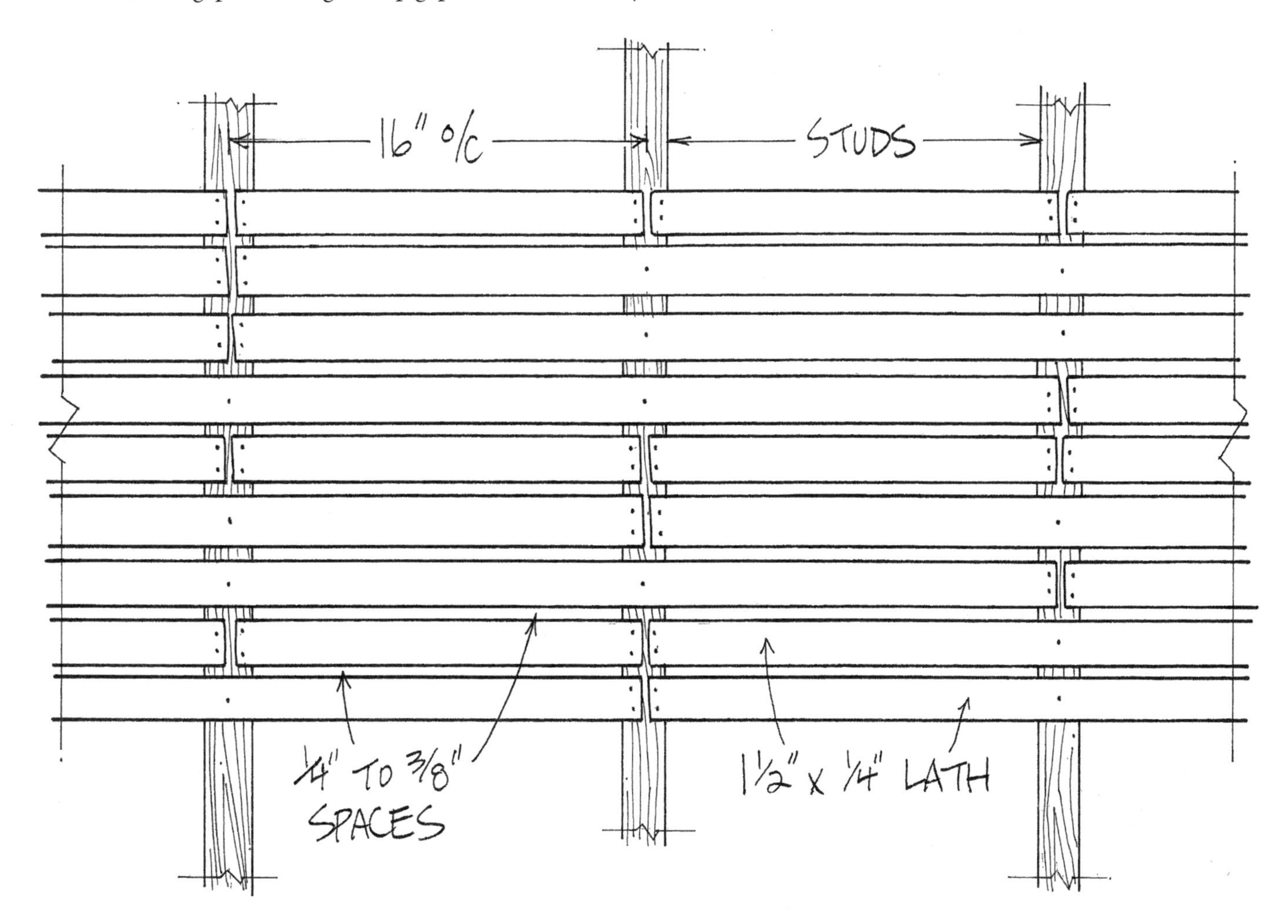

Fig. 3.4: *Stagger joints when installing lath.*

Veneer Finish Plasters

Plastering over Drywall and Cement Board

Plastering over wallboard is one of the most economical ways to introduce plaster into a conventional home, if not necessarily the most ecological. There are drywall products available that circumvent some of the negative effects of drywall. These include *paperless* drywall (which is resistant to mold), recycled drywall, and *tox-in-filtering* drywall. Nontoxic joint compounds are available. Drywall should usually be sealed before plastering, at least with wheat paste or a zero-VOC primer. A sanded adhesion coat will be called for in many, but not all cases. This can be created by adding sand to either the wheat paste or the primer coat.

Gypsum boards are available that are designed for gypsum plaster systems, but they can be used with other plasters. These are sold under the names Imperial (USG), Grand Prix (CGC), and Kal-Core (National Gypsum).

There are other types of wallboard that can be used as a substrate for plaster, including cement board and magnesium board. These may be good options for lime plasters that will retain moisture for an extended period. It's important to note that these wallboards are harder and less flexible than drywall, so they should be mounted with a small gap at each joint to accommodate expansion and movement. This gap should be filled and allowed to dry before taping and mudding.

Plastering over Drywall

By Errol Towers

Fifteen years ago, I started looking for alternatives to drywall and painted walls. This idea was a little crazy to some of my friends because, after all, I had been a drywall and painting contractor for almost 20 years. "Why would you want to throw away your livelihood?" they asked. The truth was that conventional paints were making me sick, and I still needed to make a living. I kept searching until I came across some information on clay as a plaster. A bit skeptical, I thought I would give it a try. There was only one problem. All of the walls in my house were painted drywall. Out to the garage I went. I had plenty of sheets of drywall and buckets of mud galore. Naturally, I hung a few sheets and taped them with some lightweight setting-type compound. Trust me when I say "It failed miserably." I could see the taped joints through the plaster and it just wasn't sticking. My plaster turned to dust as I ran my hand over it.

Clay, and lime plasters, as my searching continued, looked so beautiful on cob, adobe, and masonry. There had to be a way to use natural plaster on drywall. Drywall: the ubiquitous "I don't know what else to use" interior wall sheathing. In North America, it is literally the only thing we make walls out of. Back to square one I went.

Let's consider new walls for a minute. Drywall seams need to be taped. This can be done with either paper or mesh. Laying a thin coat of ready-mixed joint compound over the joints and pressing paper tape into it is the most common method. Once the tape is on, the excess compound is pressed out and smoothed with a standard taping knife and allowed to dry. Up to three more coats of compound can be smoothed over the joints, but with ever-increasing-sized taping knifes or trowels for each coat. Let dry overnight and sand between coats as needed. This works pretty well and is probably best if you are new at taping.

Mesh tape comes in self-adhering rolls and eliminates the wait time for the otherwise paper-taped seams to dry. The next step would be to apply a coat of regular setting-type compound to the mesh tape. Resist the urge to use lightweight setting compound here. The moisture from your natural plaster can turn lightweight compound to mush, but more on that in a minute. Never use ready-mixed compound as a first coat over mesh. It will hairline crack almost certainly. Regular setting compound sets very hard and is not easily sanded, so work cleanly. A second coat, this time with ready mix, can be applied if needed. ☛

Drywall preparation

Here's a short to-do list for drywall prep:

- Always use a setting-type drywall mud (e.g. Sheetrock or Durabond). Premixed (wet) drywall compounds may rehydrate under earth plasters, leading to cracking or unsightly changes in texture along joints.
- Tape and mud joints. You don't have to mud drywall screws, but another pass to level the joint is generally a good idea. Finish drywall to level 2 or level 3 (ASTM C840) depending on how thick your earth plaster coat will be.
- Avoid sanding. If spot sanding is required, wipe dust away with a wet sponge and allow the spot to dry.
- Prime with good-quality paint primer with sand in it (see Adhesion coats, below).
- If magnesium board or other hard/non-flexible wallboard alternatives are used, it is essential to leave a small gap (about ⅛") between boards. Mud the gap first and allow to dry before mudding and taping the joint. Failure to do this is likely to result in cracks at some joints.

Adhesion coats

Generally, an adhesion coat is needed before plastering over drywall, whether it be bare or painted. Adhesion coats have two functions — providing a rough surface for the plaster to bond to, and sealing the surface of the drywall. If the adhesion coat is only needed for texture, wheat

The biggest challenge is to create a smooth plane at all seams. Natural plasters can go on fairly thick and therefore hide some taping imperfections. Use this to your advantage to save time. It doesn't have to be perfect, just flat. Set a straight edge on the tapered edge seams and fill with more compound if you have to. Once you are happy with your walls, you are going to need to prime them.

Natural plasters contain copious amounts of water in them. If that water soaks lightweight, and, to a lesser degree, ready-mixed compound, then they will reactivate and become soft. Unfortunately, when it dries out, they no longer have any strength. Your seams will fail. Priming the wall will accomplish two things at this point. The first thing that it will do is to help seal out the water. Get a good-quality primer. American Clay has already done your homework for you here and tested many interior primers. Go to americanclay.com and find their list of low- and zero-VOC primers. A word of caution: do not use most PVA primers, because they allow too much moisture to pass through and into your seams. The second thing that primer does is to equalize the porosity between the taped seams and the open areas of drywall that have no compound on them. Without primer, you would likely see a color or shadowy difference in your plaster where the compound lies beneath it. We call this *ghosting.*

Most natural plasters will have little or no ability to stick to your new wall without some help. Of course, there are plenty of additives, but that is a whole discussion in itself. We need to add *tooth.* The wall needs a sharp, fine grit on it to hold the plaster. By adding fine, sharp sand in with your primer, the aggregate in your plaster has something to hold on to. Consistency in size is the key to this step. If you have to buy sand, pool sand works well. Skid-Tex is an additive for painted floors to make them slip resistant. This works very well, but it only comes in small containers. I like to use crushed limestone that I get from a local quarry. Bagged sand can be purchased from building supply stores, but isn't always very consistent. Ask to see a sample, if possible. Add just enough to the primer so that when it is rolled or brushed, it looks like 60-grit sandpaper. Previously painted drywall just needs the sanded primer applied before you can plaster.

That should do it. Just remember to use the appropriate tapes and compounds, choose a good primer, and add sharp sand for tooth, and you are ready to apply natural plasters to drywall.

Table 3.3: A Comparison of Adhesion Coats

Sanded wheat paste	Incompatible with lime plasters; totally nontoxic; doesn't seal wall completely; can be plastered over within a few hours. 5–10 wheat paste:1 sand OR 1 pre-gelatinized starch:6 sand:8 water (mix starch and sand before adding water, adjust water to suit).
Sanded paint primer	Some toxicity; seals wall completely; blocks stains; best left overnight before plastering. 8–12 parts paint primer:1 sand (roughly 2 cups of sand in a gallon of paint).
PVA glue	Low toxicity; works with all plasters. 2 glue:1 water. Use a paint roller and sprinkle dry sand into the tray as needed. May not seal wall.
Commercial bonding agents	Low toxicity; formulated for plaster bonding; important to follow directions. Examples include Plaster Bonder or Plaster Weld. May not seal wall.

paste with sand will work for earth, but not for lime plasters. To seal the wall, a good-quality paint primer with some sand added will work. A number of plaster bonding agents are available, or you can make something similar using PVA glue.

Alternatives to drywall

Various alternatives to drywall are possible, but all are more labor-intensive. Instead of framing interior partition walls, they may be built out of cob or adobe. Alternately, stud walls may have an infill of slip straw, wood chip-clay, or hemp-crete. Or lath may be attached to framed stud walls and plastered — wood lath, metal lath, or even wattle and daub are options.

Other Substrates

Concrete block walls

Most concrete block walls can be plastered directly without any intermediary, but occasionally you may find they have been treated with a sealer. To confirm that concrete blocks haven't been treated with any kind of sealer (in the factory or after construction) sprinkle some water on them — if it absorbs immediately, the concrete is ready for plaster; otherwise, you may need to sandblast to remove the sealer. An alternative would be to use metal lath, though in some cases a bonding agent will be sufficient.

Poured concrete

The main difficulty with poured concrete is that form-release agents are often used, which can inhibit bonding with the plaster. This may be removed with sandblasting, or possibly pressure washing.

Other masonry

Most masonry and brick is a good substrate for plaster without any special preparation. If masonry is loose or damaged, you may need to install lath over top to stabilize it.

Plaster Stops

A plaster stop is a piece of wood (or other material) affixed to the framing prior to plastering that dictates the depth of the plaster. It is helpful to have plaster stops installed around window openings, and at ceiling and floor junctions. The plaster then ends flush with the plaster stop, resulting in a cleaner, neater finish. A common approach to plaster stops is rough to rough/ finish to finish. This generally has the plaster stops set to the depth of the base coat, then trim and baseboard are applied and the finish plaster lands against the wood trim. This is the best system when a thin finish coat (<⅛ inch) will be used.

When thicker finish coats are used, it is common to set the plaster stops at the depth of the combined plaster coats and install trim over top of the finish coat. Unevenness in the plaster will show as a gap under the trim; one way to minimize this is to use a fairly wide nailing board as a plaster stop, so that the trim extends only ¼–½ inch over the plaster. This is a common technique for baseboard.

At the tops of walls, a ½-inch plaster stop may be installed so that ceiling drywall will butt up to it cleanly. Choose the width of the plaster stop to match your desired plaster depth (be realistic about how much you'll be able to build depth if the wall is uneven). Drywall may be

Fig. 3.5: *Plaster lands on plaster stop, and provides nailing surface for baseboard. Baseboard is installed after finish coat of plaster is complete.*

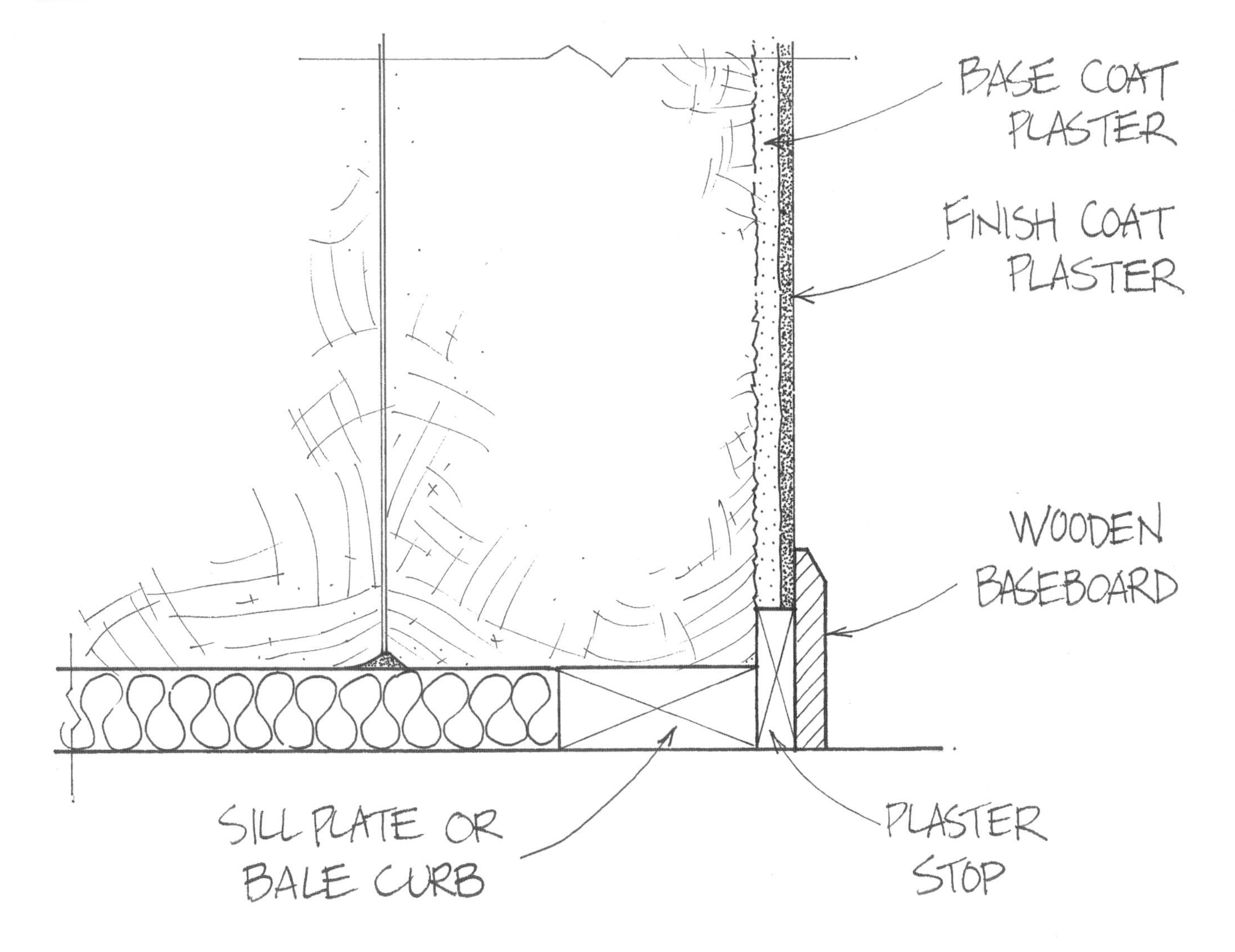

installed before the finish coat, and the finish can land on it for a very clean result. You may wish to paint the drywall with primer before plastering. Another detail is to install drywall channel before plastering — consult your drywall contractor, as channel may need to be sized wider than the drywall to fit properly.

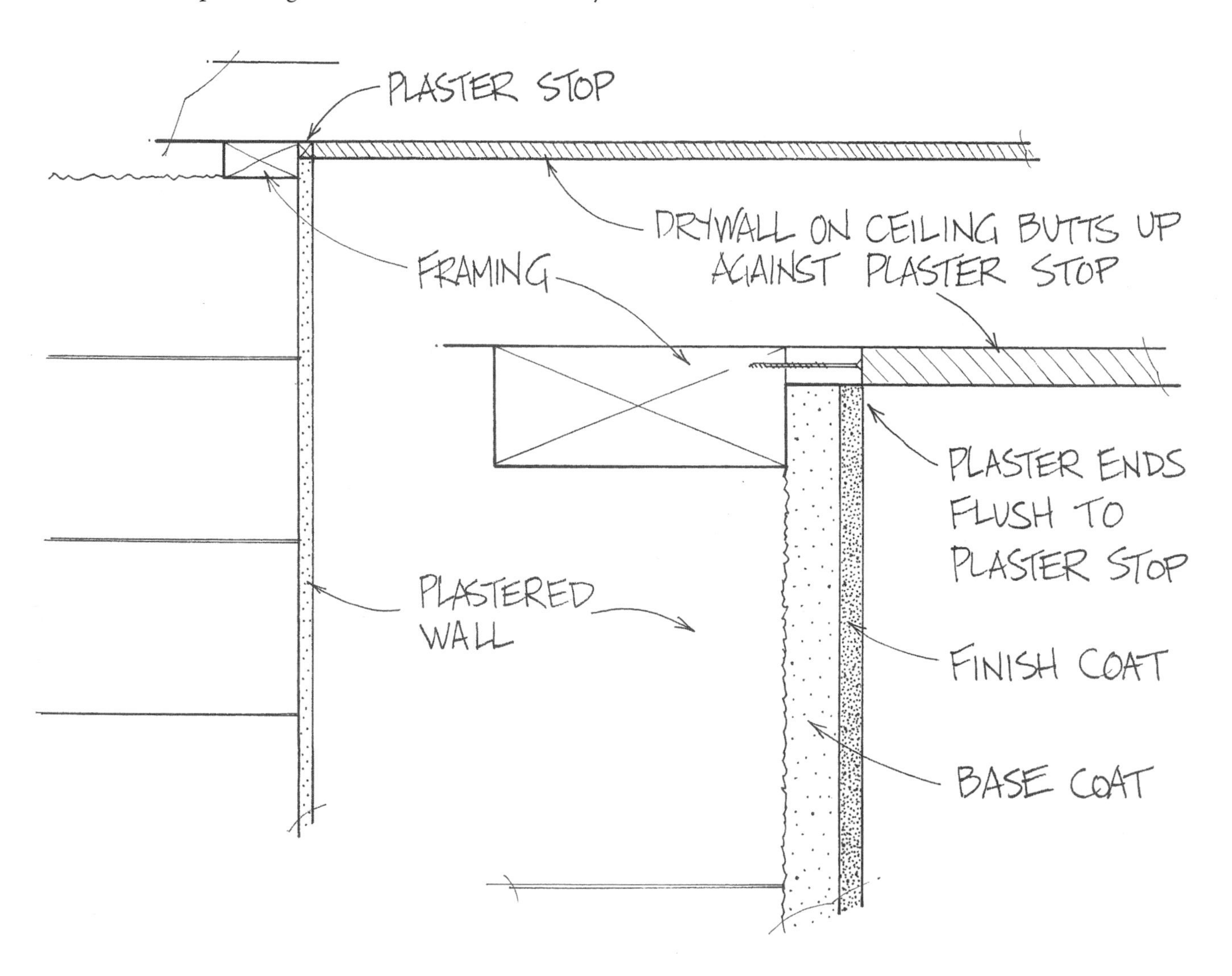

Fig. 3.6: *A plaster stop provides a clean strike-off point for plaster, and a neat edge for installing drywall.*

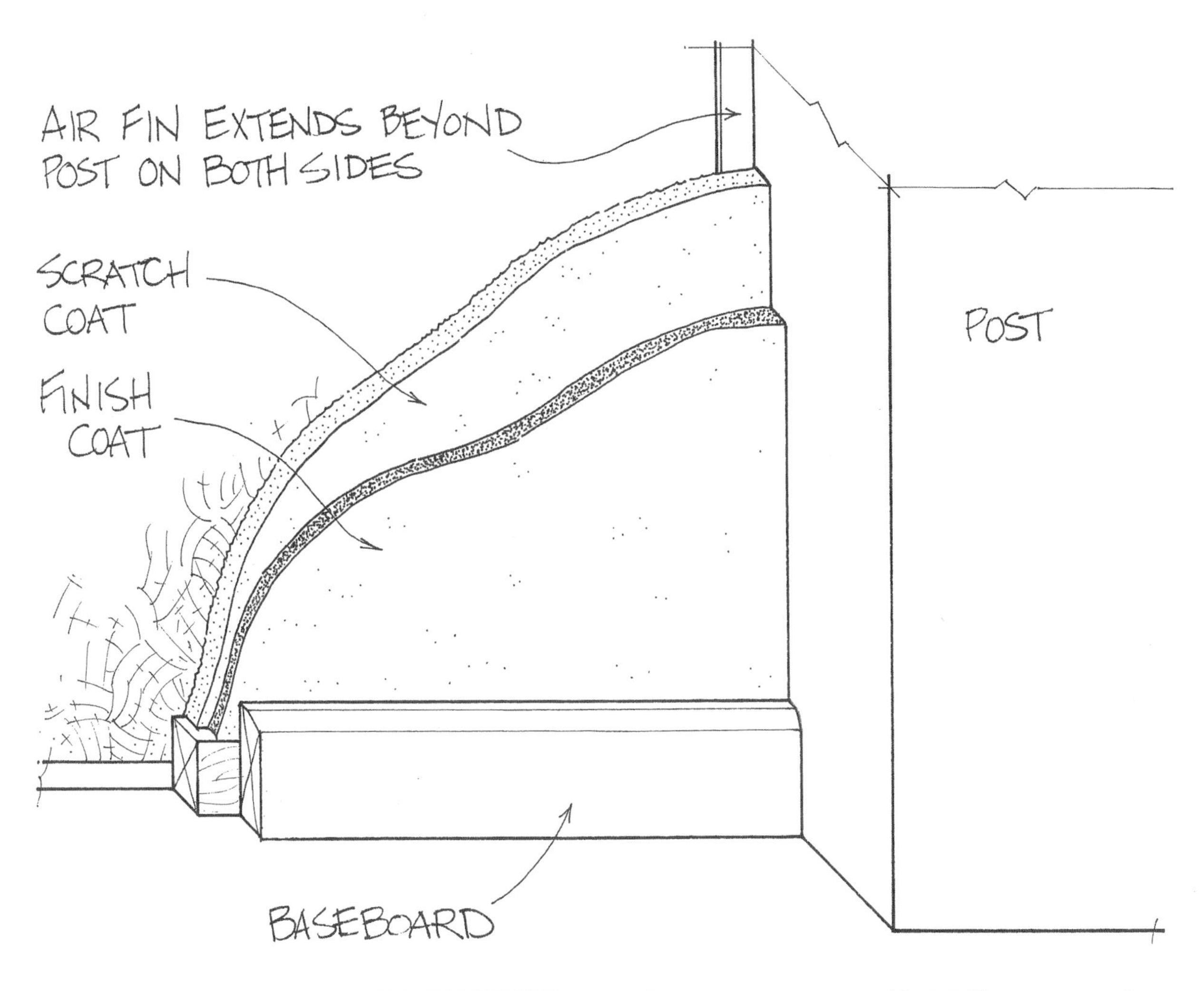

Fig. 3.7: *Plaster stops can be permanent or temporary. They offer a neat finish for ending plaster, and also allow straight materials, such as drywall, to be neatly butted up against them.*

Fig. 3.8: *Drywall channels can be installed as plaster stops for interior walls. Also note the air-sealing tape at the joints.* Credit: Michael Henry

Air Detailing

Plaster is an excellent vapor-permeable air barrier; however, every plaster edge and penetration must be detailed well in order for it to perform this function. All these details should be planned in the design phase, but sometimes plasterers need to suggest and/or implement details related to the plaster. The main tools in a plasterer's toolbox are *air fins, air membranes, air sealing tapes,* and *caulking.* Too much reliance on caulking, in particular, is bound to result in a leaky, under-performing home.

Fig. 3.9: *This timber frame structure has moved enough to break the caulking and create a sizeable air leak. An air fin extending behind the timber frame and underneath the plaster would prevent air leakage. Air sealing tape might be used to retrofit these joints.* CREDIT: MICHAEL HENRY

Air Fins and Membranes

Air barrier membranes extend underneath the plaster at wall edges and penetrations, where gaps are very likely to form. It's common to fasten vapor-permeable air barriers behind plaster stops at wall edges and run them several inches under the plaster. TYPAR housewrap is sometimes used, as are high-performance air barrier membranes produced by German companies SIGA and SOLITEX. Cheap, thin housewrap products may be far less durable than more expensive (often, imported) solutions. In the end, it can be worth bending your principles (and your pocketbook) to bring in imported materials when it results in higher performance. We are keeping our finger crossed for suitable North American products. Caulking may be needed where the air barrier meets the framing.

Air fins are commonly used in timber frame structures to create airtight joints. In the absence of an air fin, the plaster will crack or pull away from the structure at every edge, creating an opening directly into the wall. An air fin is a flexible, or preferably semi-rigid, vapor-permeable but air-impermeable material that extends behind the post or beam (with caulking, where necessary); it also extends under the plaster at least several inches. The material must not be so rigid that it simply transfers the location of the crack out to the edge of the air fin. One of the best materials for air fins is Homasote, a fiberboard made from recycled paper that is marketed as a sound barrier. It has a good balance of flexibility and rigidity, and plaster will stick to it without the need for supporting lath (a huge advantage). Another option is #30 roofing felt with lath over it.

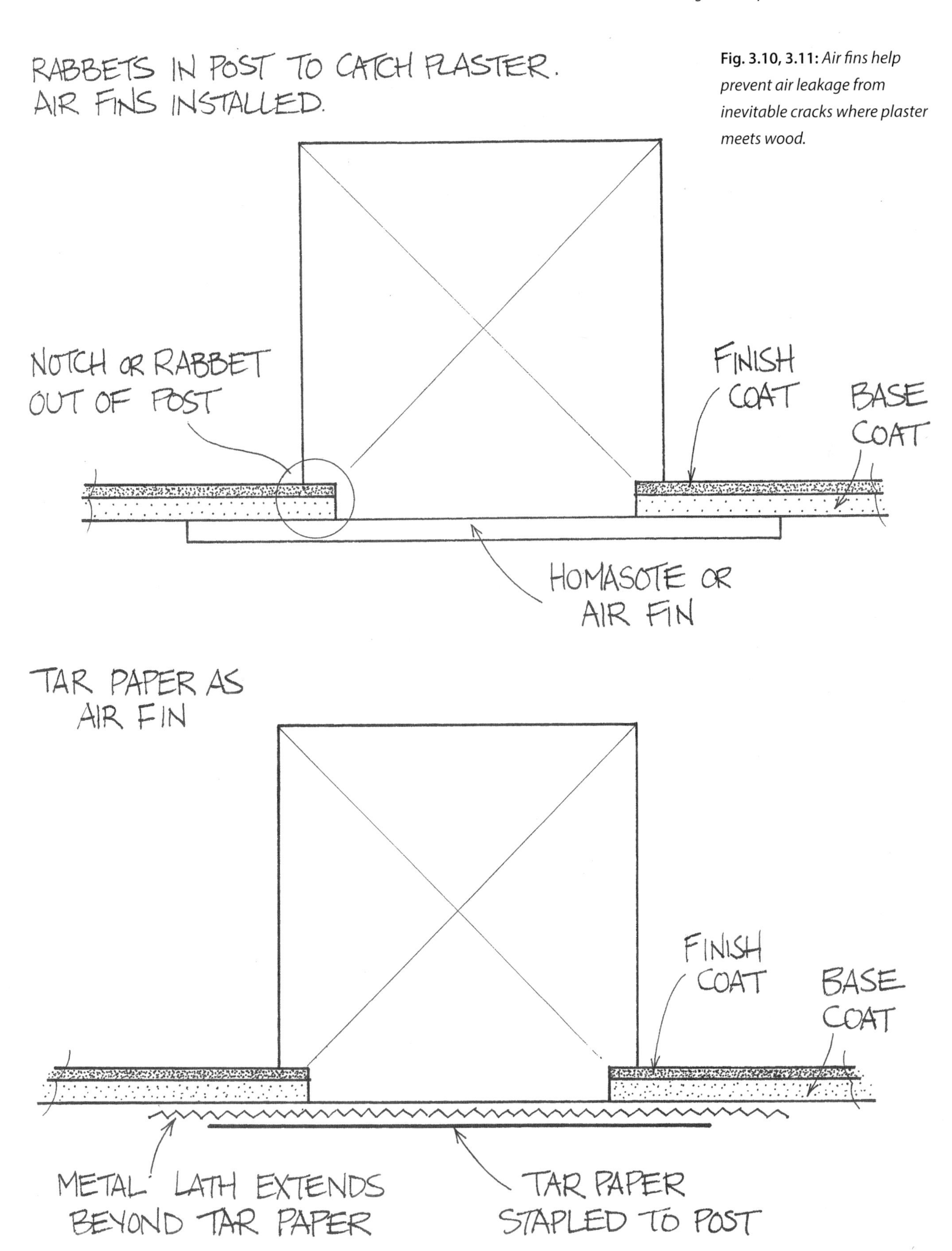

Fig. 3.10, 3.11: *Air fins help prevent air leakage from inevitable cracks where plaster meets wood.*

Air sealing tapes

In recent years, air sealing tapes have become real work-horses in detailing plaster edges. The German manufacturer SIGA makes Fentrim, a tape designed for plaster to adhere to, which is ideal for sealing tricky penetrations such as pipes or electrical boxes. Air sealing tapes can also be used between plaster coats, which could simplify details or fix mistakes.

Fig. 3.12: *Air sealing tapes used around all wall penetrations, such as this outlet box, can significantly improve the energy performance of a home.*
CREDIT: DEIRDRE MCGAHERN

A huge advantage of air sealing tapes is that they may be used to easily retrofit existing natural buildings that were not well detailed for air leakage. We recently heard of a straw bale timber frame house that was built without air barriers, and the homeowners chose to tear out the bale walls because of the very high rate of air leakage and associated moisture loading. A few rolls of tape and a couple days of work might have saved the many thousands of dollars the owner spent building a new wall system.

Caulking

Caulking has a limited — but important — role in air sealing plaster edges. Caulking can be used to seal the edges of interior and exterior finish plasters, to limit air leakage and water penetration.

Acoustical sealant is a black, gooey, and permanently sticky type of caulking that is used to seal edges of vapor barrier in most northern construction. It is also very useful in sealing cracks that will remain hidden, such as between the subfloor and a bale curb, under air fins or plaster stops, and between framing elements.

Waterproofing Penetrations

Flashing and Drip-Edges

Any penetration in the building (window, door) needs to be adequately flashed to direct water away from the building. There are many different building wraps and tapes that can be used to flash openings. Drip-edges should be installed above all openings at the beginning of plaster prep.

At the base of the wall, it is important to install metal flashing that is angled slightly away from the building. This flashing should also act as a capillary break to prevent wicking between foundation walls and plastered walls. Ask roofers for their tricks with cutting and bending flashing. Also be aware of corrosion and reactions between different metals, and make sure you're using the correct fasteners for the job.

Fig. 3.13: *Drip edge extends beyond window, to deflect water. Installed before plastering, it is covered with plaster lath to accept plaster.* CREDIT: TINA THERRIEN

Fig. 3.14, 3.15, 3.16: *Like it or not plasterers sometimes seem to get stuck putting flashing on before plastering. If you agree to do it, do it right: caulk joints of flashing, minimize fasteners, and keep penetrations high on the flashing. Pictured here is one way to cut a corner.* CREDIT: DEIRDRE MCGAHERN, MICHAEL HENRY

Clean up

Regardless of how you are mixing (plaster pump, mortar mixer, wheelbarrow), large quantities of water will be used at the mixing station for cleaning the equipment. If you don't want this area to turn into a quagmire, plan for drainage before you start mixing — it might mean digging a trench to take water away, or siting the mixing station high, so that water will drain away naturally.

Check with the owner to see where excess plaster can be dumped, and make sure that everyone on the plastering team knows about these details.

Plaster Prep Checklist

It is helpful to have one person from the plastering crew go to a straw bale jobsite prior to plastering, especially on an owner-prepped building, to make sure that all of the prepping is done to the standards you require. Actually, on an owner-prepped build, it is best to have a crew member show up at the beginning of the prepping process to set the standards, and even part-way through, to make sure that things are on track.

Moving Mud

You will have to figure out how to move plaster from the mixing station to the plasterers. If the building is tall, there will be a lot of scaffolding. Sending the plaster up to a fully decked scaffold can be tricky. Pulley systems may be permitted in your region; it is your responsibility to know what the construction safety standards are and make sure they are complied with at a jobsite. Some professional crews use an automated scaffold elevator that is incorporated into the

Table 3.4

Checklist: Prior to Plastering a Bale Wall
❑ Bales are trimmed.
❑ Walls are straightened.
❑ Exterior corners are plumb.
❑ Holes or spaces between bales in walls are tightly stuffed with straw/cob/slipstraw.
❑ All window and door openings are shaped as desired, and, if mesh is to be used, it is well installed, not loose.
❑ Wood surfaces are covered with barrier paper, then diamond lath/mesh, or burlap; air fins installed if needed.
❑ Plaster stops installed.
❑ If extra bracing required for installing cupboards, it should be affixed appropriately on wall.
❑ Drip-edge installed above windows and doors, under reinforcement mesh.
❑ Flashing should be installed before reinforcement mesh.
❑ Airtight electrical boxes properly installed and sealed with air sealing tape.
❑ Reinforcement mesh (if using) is hung inside and out (stapled or nailed taut to top plate and bottom curb and any other wooden framing).
❑ Reinforcement mesh should NOT extend beyond the bottom curb, or the top plate. It should be neatly and tightly installed and stitched, or quilted.
❑ Water is on site for mixing and cleaning equipment (preferably pressurized).
❑ Electricity is hooked up (if critical to your plastering methods; if you are using an electric compressor, you will need electricity).
❑ Scaffolding should be on site and erected prior to plastering; scaffolding should be plentiful enough to surround house (or you pay for the time that the crew takes to move the scaffolding around the house as they plaster!), and it should be properly erected according to construction safety standards.
❑ Windows and doors, finished floors, and ceilings/timbers are masked and/or tarped with quality tape.

scaffolding — an expensive, but necessary investment if you plaster tall buildings.

If you are fortunate enough to plaster only tiny houses, or single-story buildings, you may be able to pass buckets of plaster to the plasterers by hand. Buckets of plaster are extremely heavy, so make sure to fill them an appropriate amount that is manageable for all of the plasterers on the crew.

Having a couple of wheelbarrows on site is useful, to keep the mud moving.

Scaffolding and Ladders/Steps

A three-step ladder allows access to the tops of most single-story walls, and two three-steps can be used with a scaffold plank to allow plasterers to easily work side by side. Avoid working off the top step of a three-step (use a step ladder, or preferably scaffolding).

Working off scaffold is almost always the best option, except for low walls where a three-step may be more appropriate. Below are a few tips for scaffold setup, but learn the proper procedure for legal, safe scaffold setup in your jurisdiction, then follow it.

- Always use screw jacks for outdoor work. Wheels are often preferable for indoor scaffolding, but make sure all wheels are locked when in use.
- Start where the ground is highest and set the screw jacks to the minimum height.

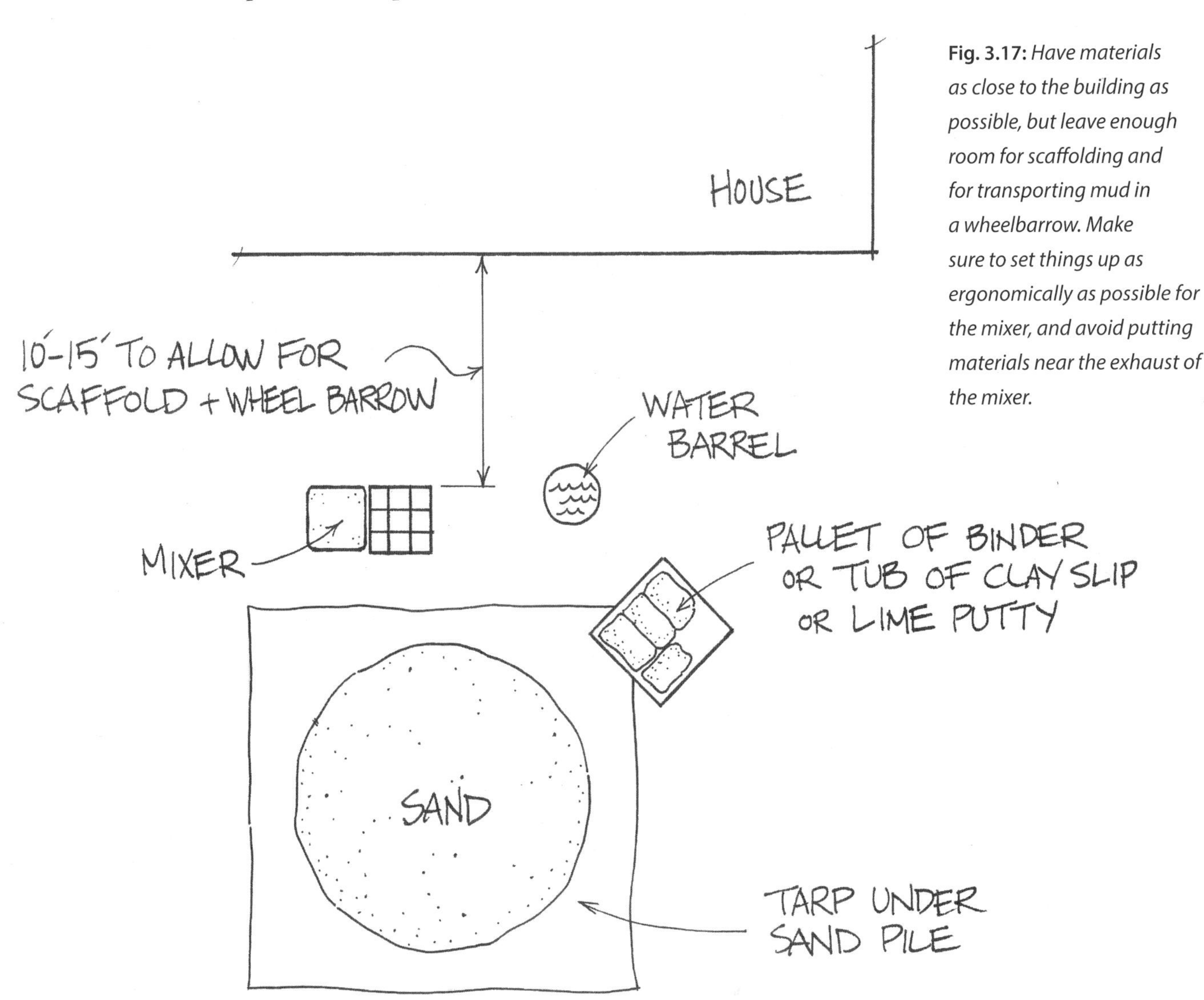

Fig. 3.17: *Have materials as close to the building as possible, but leave enough room for scaffolding and for transporting mud in a wheelbarrow. Make sure to set things up as ergonomically as possible for the mixer, and avoid putting materials near the exhaust of the mixer.*

- Always properly level scaffold (a small magnetic level works well). Level the first scaffold frame, then use a plank to level to the next frame, and check that frames are plumb.
- Avoid stacking anything under legs — if the ground drops off more than the screw jacks can accommodate, consider using 36-inch high frames to help with leveling. If a support structure is necessary, have it built properly by a carpenter on site.
- Anchor to the building if more than two lifts of scaffold are used. Sometimes anchor points will impede plastering and may need to be removed and plastered just before a scaffold is dismantled.
- Know the law and follow it. This will mean using guard rails and having ladders for access.

Protecting the Work

Plaster must be protected from heat, cold, wind, and rain during — and usually for some time after — plastering. For interior plasters, this may mean ensuring there's adequate heat and ventilation for drying. For exterior plasters, this basically means a lot of tarps.

The most common approach to tarping is the *permanent tent method,* whereby tarps are hung from the outer soffit area at the top and attached to the ground below for the duration.

The other approach to tarping is the *roll-up method,* wherein the tarps are weighted at the base with a small piece of strapping (usually 1×4 or 2×2); when the weather is nice, the tarps can be rolled up and temporarily anchored with screws. While it can be pleasant to work with the tarps up, the roll-up method typically requires more labor (to regularly roll them up and down).

Tips for setting up tarps:

- To anchor the top of a tarp, roll it 2–3 times around strapping and screw it to the underside of the soffit or bottom of the fascia board — *not* to the front face of the fascia, or water will come in when it rains.
- Plan the direction of roll so the tarp hangs behind the board, not in front; if you screw through the free-hanging tarp, you'll lose your screw heads in the first wind, which makes removal much harder.

Fig. 3.18: *Tarps extend beyond scaffolding, protecting the walls (and workers) from the elements.*
CREDIT: TINA THERRIEN

- Don't weight the bottom of the tarp to hold it down — a gust of wind can turn very heavy objects into battering rams or missiles.
- Wrap the bottom of the tarp around strapping (same as top) and screw into 2×2 stakes that have been pounded well into the ground.
- At vertical joints where tarps meet, wrap each tarp edge 1–2 wraps around strapping and then screw the strapping together.
- Where the tarp rubs on the top of scaffold, leave the pins on the top frames, but put yogurt containers over the pins to prevent them from making holes in the tarp.
- This setup uses a lot of 1×4 strapping (2.5–3 linear times the perimeter of the house), but the system will last the duration of the job, even through storms. Loose lumber attached to tarps can break windows or cause injuries — everything must be well anchored and tight.

Masking

Plastering is unbelievably messy, and masking is an essential part of it unless there will be no finished surfaces — but even then, rudimentary masking can speed cleanup. It's very important to mask wood surfaces well. If possible, finish all wood surfaces before plastering, as raw wood will easily stain when lime or earth plasters touch it (some, such as cherry and oak, are especially vulnerable to dark staining from lime). Oiling wood won't necessarily protect it from staining, depending on the type of wood.

Everyone masks a little differently, but our system is fairly simple and effective, so we'll go over it for you. It's basically a three-step process:

- Tape the edge with masking tape. Apply this tape carefully.
- Stick on another layer of tape covering about half the masking tape. Leave the edge of this tape loose.
- Slide lightweight masking plastic under the raised edge of the tape, press it down.

Whether the lightweight plastic is covering a window, door, or post or beam, the masking is generally pretty similar. Start by anchoring it at the top so that gravity works with you, then tape and anchor the sides and bottom. We usually use masking tape for layer 1 and technical tape (used for sealing vapor barrier) for layer 2. This is slightly overkill, but it saves having the plastic fall off as soon as some plaster falls on it or the wind blows.

Another way to mask is to use stucco tape to replace both tape layers — apply it along the edge and leave the lip hanging loose to catch the plastic. However, some stucco tapes are not sticky enough and tend to let go, so pick up a roll (if you can find it) and try it out before you commit to using a lot of it.

Finished floors need to be masked extremely well. Kraft paper and other floor masking materials are available, but you should first lay down a layer of plastic to protect the floor (plastic alone is far too slippery to be used on its own). Tape joints and edges well.

Cutting vs. pulling

The first choice you need to make before you start masking is will you cut the tape after the plaster has cured, or will you pull all the tape right off after applying the plaster and then touch up any damaged areas? Both have definite advantages. In general, cutting the tape is faster, but there's a risk that it will look less clean and professional if not done very carefully. Pulling the tape takes longer, but you will be able to do any touch-ups before you leave the jobsite — also the tape line can help define the plaster edge and keep it straight.

If you're going to cut the tape, you can run the tape well under where the finish plaster will

land. The main thing you want to do is keep your tape layer thin and free of any bubbles or folds — these will show and tend to make your plaster job look amateur. Also avoid having more than one layer of tape run under the plaster. During plastering, keep a nice clean sharp plaster edge that you can follow later with a knife. When cutting the masking after the plaster has cured, have plenty of good-quality blades handy, and snap them regularly — a very clean cut is essential to a good result.

Scheduling

Even when you are able to be flexible enough to adapt to a changing construction schedule, be realistic with yourself and with clients. Plasters require certain temperatures and conditions, and it simply won't work to force things.

Plan to apply earth plasters in the heat of summer, lime in early autumn. In cold climates, don't plaster in winter, except for thin interior finish coats. Plan construction accordingly. Natural materials have more benefits than synthetic materials, but also more limitations. Understand them and work with them.

Fig. 3.19, 3.20: *If you're going to peel the tape, do it before the plaster is leather hard — while it is still easily repaired.* CREDIT: DEIRDRE MCGAHERN

Chapter 4

Mixing and Application

Mixing

THERE ARE MANY WAYS to mix plaster (reviewed below), but regardless of which method you select, it's essential to set things up ergonomically for the person doing the mixing. They will move tons (literally, *tons*) of materials over the course of the job, and if they have to walk extra steps to go and shovel sand, or get bagged materials, it will be exhausting.

It is critical with any type of mixing to wear appropriate safety gear — which generally includes safety goggles, sleeves, pants, gloves, and always, a proper respirator (see Chapter 1). Respirators are especially important when you sieve materials, which you may need to do if your sand isn't well graded. (Diamond lath can make a serviceable, roughly ¼-inch sieve.)

We usually work *by volume* when mixing, and we like to use buckets instead of shovels for measuring sand in order to keep our ratios in check. If you really want to use a shovel, carefully measure how many shovels fill a given volume, and repeat it a number of times to ensure consistency. Recheck this from time to time; even experienced mixers may vary a little over time.

Methods of Mixing

Hand mixing

Some of the most satisfying mixing is done by hand — either in a wheelbarrow, in a wooden box, or on the ground. In many parts of the world, this is still the practice. There aren't any loud machines to contend with, or fuel to purchase, or jam-ups from an unwieldy load. When mixing by hand, it is easier to *feel* the mix. Like baking, just the right amount of water is needed — not less, not more; so a good mixer, like a good baker, will have a feel for how much is too much, and can tell by looking at a mix if it is ready.

If mixing by hand, you can either mix directly onto a sheet of plywood, or you can make a mortar box (a sturdy, shallow, rectangular wooden box for mixing mortar). Start by mixing the sand with the binders and pozzolans, and mix them well, in the same way you would mix flour, baking powder, and salt together for a cake. You can use a mortar hoe or a shovel. Once they are well combined, make a well in the middle of the dry materials and put in half of the water. Gradually draw the dry materials into the water, mixing thoroughly. Add more water as needed, and at this point, you can add any fibers you have, making sure to separate them well. Add pigments that have been premixed with water.

If you are making lime putty, make sure to remove any standing water on top of the putty (but keep that water, because it can be used to make limewash). Mix up the putty with a drill and paddle (this is called *knocking up* the plaster), fluffing it up. Once it is knocked up, put about half of it onto the mortar board, or into a mortar box, add some of the sand (about ¼), and incorporate well. Once they are well mixed, add the rest of the putty, mix together, and gradually add the rest of the sand until it is well mixed. If you feel you have to add water, add it sparingly.

Hand mixing in a wheelbarrow

Mixing in a wheelbarrow would be done in the same order and method as when mixing on the ground, using a mortar hoe as the tool. Mixing directly into a wheelbarrow is a handy way to mix, as you can then wheel the finished plaster directly to where it is needed, then dump it into a container or into buckets. The downside is, you may not be able to fit a whole mix in the wheelbarrow.

Mechanized mixers

If you decide to use a mixer, use the right mixer for the job. For large batches, this will likely be a mortar mixer; for small batches, a drill or paddle mixer may be more appropriate. With any motorized mixer, never, ever stick your hand or even a tool into it while it is running — always turn it off before reaching into it. Prior to using the mixer, ensure that it has been checked over. Is it fueled, oiled, greased, and clean? Make sure that everyone on the jobsite understands how to shut down a mixer in case of emergency.

Fig. 4.1: *A mortar mixer is the best way to mix many body coat plasters. Always leave the protective cage in place while the mixer is running.* Credit: Michael Henry

Mortar mixers

In a mortar mixer, upright blades cut through plaster ingredients without splashing them over the sides, mixing, rather than tumbling the ingredients. These mixers are commonly used in North America for mixing cement, lime, and sometimes clay plasters. There are newer, more ergonomic mortar mixers, whereby you can load bags of material at waist height (vs. over your shoulders). If the mortar mixer is too high for the person mixing, it is advisable to build a platform for them to stand on while loading the machine, to make it more manageable.

Standard mixing order when using a mortar mixer:

- Most of the water (80–90%)
- Half the sand
- Binder
- Fiber
- Other additives
- Remaining sand
- Adjust water

You will have to have made a mix already to determine the quantity of water needed. This amount will vary depending on the weather, how damp the sand pile is, etc., so be sparing. When we use a mortar mixer, we put in most of the water to start with, half of the sand, and then the binder. We mix these ingredients until everything is well coated. At that point, we add the fiber (separating it into strands to avoid clumps) and any other additives. You will want to take a peek into the mixer to see how it looks — and sounds. A good mixer (the human mixer) will know if it is too dry, too wet, or just right. If it is too dry, and you add the next batch of sand, it can put too much demand on the mixer, which can jam the paddles, or possibly break the belt. If it looks fine, add the remaining sand, and adjust the water if necessary. If using lime putty, knock it up first, then add it to the mixer. Add the sand gradually until the two are well mixed. As always, if you have to add water, add a minimal amount. Add fibers at this point, right before using the mix. Follow this same lime putty procedure for any of the other mechanized mixers. This is the mixing order for dry binder. When mixing with clay slip it will need to be added first to introduce enough water for the mix. Lime putty would typically be added after half the sand (unless most of the mix water was blended with the lime putty).

Cement mixers

Cement, or concrete mixers, have a barrel that turns. Designed for concrete, they aren't the ideal choice for making plaster, but we've used them on occasion. It usually takes longer, and sometimes materials get bound on the sides, rather than mixing together. When using a cement mixer for plaster, put all of the dry ingredients into the mixer first, to allow them to mix together. Add the water gradually, occasionally stopping the machine to get materials off of the sides of the barrel (make sure the drum isn't turning if you are going to reach inside). You may also have to tilt the barrel to get materials to fall off of the sides and then mix together. In our experience, you need to use more water in a cement mixer in order to fully mix the plaster.

Drill and mixing paddle

A common way to mix small batches of finish plaster is to use a drill with a mixing paddle and a mixing tub or bucket. Use a high-quality mixing drill with a dedicated mixing paddle. Think of ergonomics here. Rather than requiring the (human) mixer to stoop over a bucket, elevate the bucket or use a paddle with a longer shaft. Try to keep your drill clean — its life can be extended by protecting it from both dust and wet clay.

There are a lot of variations on this basic setup. We have even seen people use drill presses set up over barrels to mix plaster, and then the barrel is dumped into a wheelbarrow once mixed thoroughly.

A variety of mixing orders are possible when using a paddle mixer — when following the typical mixing order (putting half the sand in after the water), it can be hard to get the sand stirred up into the mix. We like to use a two-tub setup: in one, place the precise amount of water (if known), in the other premix the dry ingredients (add binder before sand for easier mixing). Add dry to wet all at once, then mix well.

Dry clay can be mixed into a slip, and then the remaining ingredients can be added. To do this, add the clay to the water and let it sink in before mixing. This reduces dust and improves mixing, while only adding a few minutes to the mix time (use the time to measure other ingredients).

Pan (paddle) mixer

Paddle mixers are more common in Europe than in North America for mixing lime putties and plasters. They operate somewhat like a large bakery mixer, with rotating blades inside, and sometimes scrapers. With a pan mixer, you add dry ingredients first and let them mix really well together before adding water. If using lime putty, put the putty in first and let it mix for a couple of minutes before slowly adding the sand. The mix is generally ready after churning for about 10 minutes. Generally speaking, there is a trap door at the side or bottom of the pan to let the plaster out once it is mixed. This type of mixer is more commonly used for lime plaster, although it can also be used for earth plaster.

Application

Plastering is a skilled trade that you can spend decades mastering. Videos can help, but look for opportunities to learn from skilled plasterers, either in a workshop setting, or by volunteering with a professional. As a starting point, we review some tools and techniques below.

Trowels and Floats

Always buy good-quality trowels. They will last longer and will usually work much better for you from day one. It's often worth sighting down the line of a trowel to see if it's straight — especially for pool trowels. If you don't know where to find good trowels, start with the best masonry supply store you know of, or you might try a good hardware store in your local "Little Italy," or ask local masons and plasterers where they shop. For some specialty trowels, you may need to shop online.

Smaller trowels can be easier on the body, but it's easier to level a wall with larger trowels. As a general rule, use as large a trowel as you feel comfortable with, since it will speed the work and make a flatter wall. Be realistic; don't injure yourself. You may wish to apply mud with a smaller trowel and then switch to larger trowels for leveling and finishing.

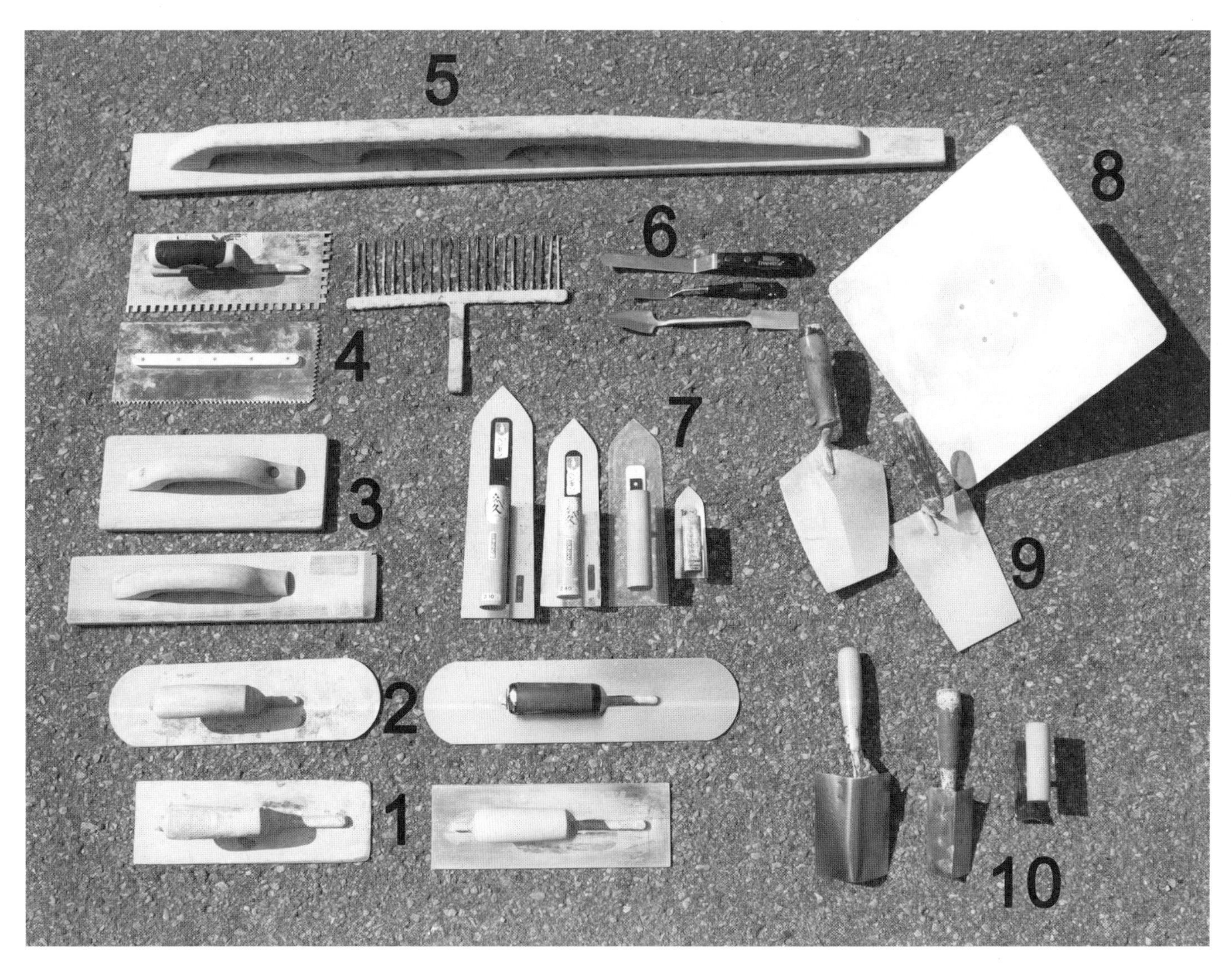

Fig. 4.2: 1 *rectangular floats;* **2** *pool trowels;* **3** *wood floats;* **4** *scratch trowels and scarifier;* **5** *darby;* **6** *palette knives and leaf and square;* **7** *Japanese trowels;* **8** *hawk;* **9** *bucket trowels;* **10** *corner trowels.* Credit: Michael Henry

The material trowels are made from, and their shape, are both important. Generally speaking, hard rectangular trowels are used for leveling; more flexible and sometimes rounded trowels are used for finishing. Also, as a rule soft "grabby" materials such as wood, or porous metals, are used for leveling, while hard slippery materials (such as stainless steel) are used for the final finish.

Wood floats

Wood floats (trowels) are often used for leveling earth and lime base coats. Generally, the plaster is applied using a steel trowel (or by hand) and then *floated* (troweled, or rubbed) with wood to level it. The texture left by the wood float helps the plasterer see low spots that can then be filled; also, the wood tends to drag the plaster around and help level it.

Darbies

A darby is a long rectangular flat trowel, ranging from 3–6 feet long, with one or two handles on it. It is used to level large sections of basecoat. Wooden darbies function somewhat like wood floats, dragging some material as they are floated over the plaster surface. Aluminum darbies don't move material as much, and they often have one side ribbed to help show low spots on the wall. In either case, material is added into the low spots before the wall is floated again with the darby (a darby can be seen in Figures 4.2 and 7.1).

Steel trowels

Steel trowels are commonly used at every stage of plastering, but they are especially important for finishing. Stainless steel is useful when working with earth plasters because earth plasters tend to cause rust on steel tools. Stainless steel is very hard and smooth, which makes it great for finishing, but less ideal for leveling a wall (they are usually adequate, however).

Rectangular trowels

A good rectangular trowel can be used for application and finishing of most types of plaster. Over time, the corners of the trowel will wear and become slightly rounded; this is very desirable for finishing, and a well-used rectangular trowel is often far more valued by its owner than a new one. For this reason, it is recommended to use an older, well-worn trowel for finishing, and a newer trowel for application, so that it can be worn in and improve as a finish trowel.

Pool trowels

Pool (or swimming pool) trowels are flat-bladed, have rounded ends, and are useful on curved surfaces, such as uneven straw bale walls, or other undulating natural walls. If pool trowels are used for finishing flat walls, they should only be used for a final pass after the wall has been well-leveled and smoothed using a rectangular trowel.

Japanese trowels

Beautifully handcrafted, Japanese trowels are held differently than American trowels, and they can be more comfortable in the hand.

A huge variety of Japanese trowels are available — used for everything from application of plaster to fine finishing. These have a large variety of shapes and materials — stiff trowels made from softer, more porous steel are used for application because they have the ability to float and level a wall. Hard steel and stainless steel flexible trowels are used for finishing; this type of thin finishing trowel is the most common Japanese trowel available in North America, and it is very useful for burnishing earth and lime walls, or for touch-up work. The pointed-tip Japanese trowel is ideal for getting into tricky corners around braces on a timber frame, and the mini trowels have a multitude of uses.

In *Japan's Clay Walls,* Emily Reynolds describes how to hold a Japanese trowel. She says that a lot of the control of the trowel comes from the back of your fingers pushing against the back side of the trowel.

Plastic trowels

Plastic trowels are often used for burnishing (polishing) pigmented plasters because they don't leave dark burnish marks, which steel trowels may do. Both Western style and Japanese plastic trowels are available, with Japanese being the most commonly used for burnishing. These are available from suppliers of Japanese trowels, American Clay, or other online retailers (see Appendix 2 for suppliers).

Special purpose trowels

Small tools

Leaf and square/trowel and square

This is one of our favorite tools; we like to have some variation of a leaf and square in a side pocket or a nearby toolkit at all times. We use

Fig. 4.3: *For doing half-round columns between windows, a piece of round metal duct can do most of the shaping and smoothing. All that's needed is a pass over it with a Japanese finish trowel once the mud has firmed up.*
Credit: Michael Henry

them to get into hard-to-reach places or for minor touch-ups. They are often available at good masonry supply stores or anywhere that quality trowels are available.

Palette knives

It's worth having a few artist's palette knives on hand for finish touch-ups and to get into the smallest of spaces. These are available at many art stores — get the steel ones, not plastic.

Corner trowels

It's delightful to have the right corner trowel for the job. Avoid drywall corner trowels, which flare out beyond 90 degrees — trying to use one for plastering is a lesson in frustration. Good masonry supply stores may carry small corner trowels for stucco work, etc. that are a true 90 degrees. At a minimum, it is generally worth having one square inside corner and one rounded. For larger radius curves, pieces of ABS drain pipe in various sizes are often littered around a jobsite.

Scratch trowels

A scratch trowel is used in between coats of plaster while the base coat is still green (workable). The surface of the wall is scratched horizontally to provide a mechanical bond with the next coat. Scratch trowels may be as simple as a notched tile scratch trowel, but for uneven walls, these quickly become frustrating. A pool trowel may be turned into a scratch trowel by notching it with a grinder, which works well on bale walls. A plaster *scarifier* looks like a giant comb, with flat, flexible tines. These get easily into curved and undulating surfaces, and they produce a relatively deep scratch. The only disadvantage is that the scratch can be *too* deep — if the finish will be a thin coat, or the finish plaster is applied quite wet, there's a risk of the scratch showing through to the finish.

DIY trowels

Sometimes, you just don't have the trowel you need, or it doesn't exist. You can try to make it. The simplest and most useful DIY trowel is probably the yogurt lid. Cut a circle out of any flexible plastic lid, and you have a trowel that can be used for smoothing and burnishing earth or lime plasters. It's especially useful for niches, sculptures, etc. The plastic lid can be cut to whatever shape best fits what you need, but avoid sharp edges or square corners that will dig in to your plaster.

Hawks

Hawks are used for holding small amounts of mud, particularly for application of finer finish plasters. Plaster is scooped onto the hawk using a bucket trowel, and then the hawk is tipped mostly sideways so that mud can be scraped off with a trowel. Each time some mud is removed, the hawk is rotated a quarter turn so that the weight remains balanced.

Bucket trowels and scoops

Bucket trowels have a flattened end so that they can easily scoop mud from the tub. If you are working with earth plasters, stainless steel bucket trowels are highly recommended to prevent rust from contaminating the plaster.

Other tools

Misters, sponges, and masonry brushes

Misters are critical for dampening the substrate before plastering and for hydrating lime-based plasters during curing. Plasterers will need a misting hose attachment and a good pump-action sprayer, as well as a couple of cheap hand-misting bottles. For hose attachments, the go-to in Ontario has long been the fogging nozzle available affordably from Lee Valley Tools (Canadian-based, but ships to U.S. and internationally). It provides a fine mist and has proven nearly indestructible, but it needs a separate shut-off. Pump-up misters, hand-held or backpack, are essential equipment but are notorious for their short lifespan on the jobsite, seemingly regardless of brand. The best solution is to buy any decent-quality sprayer and then be obsessive about using clean water and avoiding all sources of contamination when filling it.

Mud pans and buckets

You'll need buckets and containers of every description for mixing, moving, and applying mud.

Measuring buckets are inexpensive and come in a variety of sizes from under one quart to five gallons (<1 L to 19 L) — keep an assortment on hand. They can be hard to find sometimes; look for them where you buy your trowels.

Mud buckets, or buckets for moving mix, will preferably be small (3 gallons or less) and flexible, so that they are easy to clean and won't break when dropped. Rinse mud buckets at the beginning of the day, before the first use, to prevent the mix from sticking to them.

Rubber-type buckets and containers work very well for plaster, and they are relatively indestructible. This is especially important for large mixing tubs, which otherwise have a short lifespan. Look for rubber livestock feed containers at your local farm coop — often, they are better and/or cheaper than what you might find at a masonry supply or hardware store. We most often use Tuff Stuff bins of various sizes; for mixing, look for the 13-gallon or 17-gallon heavy-duty drums.

Mortar pans can be used for small-batch mixing or to work out of for plaster application. An inexpensive alternative can be small rubber feed/water tubs.

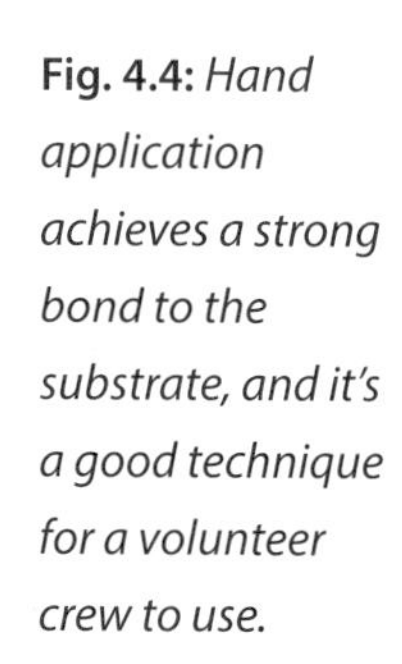

Fig. 4.4: *Hand application achieves a strong bond to the substrate, and it's a good technique for a volunteer crew to use.*
CREDIT: TINA THERRIEN

Application Techniques

Fundamentals of Base Coats

- Base coats may be applied from the top down or bottom up. We usually work top-down, but starting at the base of the wall is common with hand-applied earth plasters.
- On most natural substrates, base coats must be applied in two passes to get a good bond.
- The first pass pushes a thin bonding layer of mud into the substrate with lots of pressure. When applying mud by hand, push hard with your palm, and work it in with your fingers in trouble spots. For trowel application, angling the trowel around 30 degrees from the wall achieves a thin coat and produces a lot of pressure along the trailing edge of the trowel, squeezing the mud into the wall.
- Once the bonding layer has been applied over a small section (a few square feet, or whatever you can easily reach without moving), go back over it, applying mud to the desired thickness either by hand or by trowel.

Fig. 4.5: *A base coat is applied with a trowel in two passes: initially the mud is pushed into the wall with a steep trowel angle, then it is built up and leveled with a shallow trowel angle and less pressure.*
CREDIT: MICHAEL HENRY

- Once a large area has been applied, it should be further leveled with a wood float, a darby, or several quick passes with a rectangular steel trowel. The quality of the finish coat is dependent on having a well-leveled base coat to work over top of.
- If a wall needs to be perfectly flat and plumb (as behind kitchen cabinets) plumb dots may be used (the technique is described below).

Using plumb dots to level a wall:

- Check that your room is square.
- Draw a straight line on the floor marking the desired finish line of the plaster coat.
- Put a dab of plaster near one side of the wall and about 12" (30.5 cm) from the ceiling.
- Embed a small piece of wood (the *dot*) no thicker than the desired plaster depth into this dab.
- Repeat these steps lower on the wall, about 12" above the floor.
- Use a level to check plumb between the dots and the line on the floor — if they aren't plumb, use plaster to adjust the dots.
- Put two more dots on the far side of the wall (top and bottom), and plumb them as well.
- Let the plaster set up a little so the dots won't easily shift.
- Fill in a horizontal line of plaster between your upper dots, and another between your lower dots.
- Use a straight edge to screed each line perfectly flat, adding material to fill hollows if needed.
- Once the screed lines of plaster have firmed up, use them to level the rest of the wall.

Machine application (spraying or power-trowel)

Spraying plaster can get a lot of material onto the wall very quickly — but you need an experienced crew available to level and finish it. Fiber-rich plasters are usually not appropriate for spraying, but if you want to do this, the plaster may need to be mixed a little wetter than usual. Mortarsprayer.com offers a variety of sprayers that hook up to air compressors.

Plaster still needs to be moved to the sprayer and poured into the hopper, and a crew needs to be finishing behind it — but given the right plaster, spraying can certainly speed up a job. Everyone on site needs to wear breathing protection for the duration of the plastering.

For many years we used a plaster pump and a slotted trowel attached to the hose (the *power trowel*). It was a great way to get a lot of mud on the wall without having to bucket it, and without the downsides of spraying. The cost and maintenance of a plaster pump is a significant investment, though, and many fibrous plasters simply can't be pumped.

Floating, compression, and scouring

Floating, compression, and scouring all take place with a wood float, and the timing is the main difference. Floating generally happens when the plaster has set up a little but is workable, and its purpose is to flatten the wall. Some plasterers like to compress their base coats — once the plaster is at or near leather hard, they pass over it with a wooden float pushing hard. The plaster can get leveled by a scouring action, and the edge of the trowel will likely scrape off some of the high spots. The purpose of the compression is to add strength, and the scouring is to flatten the wall, but for many plasters, it is optional. If you're looking for a perfect wall, floating and/or scouring can help — both techniques leave a rough enough surface for fairly thin (up to 3⁄16") finish coats. If your finish coat will be thicker than this, a scratch trowel or comb might work better.

Fundamentals of Finish Coats

- Clean up the edges of the base coat and scrape off burrs or small ridges before starting.

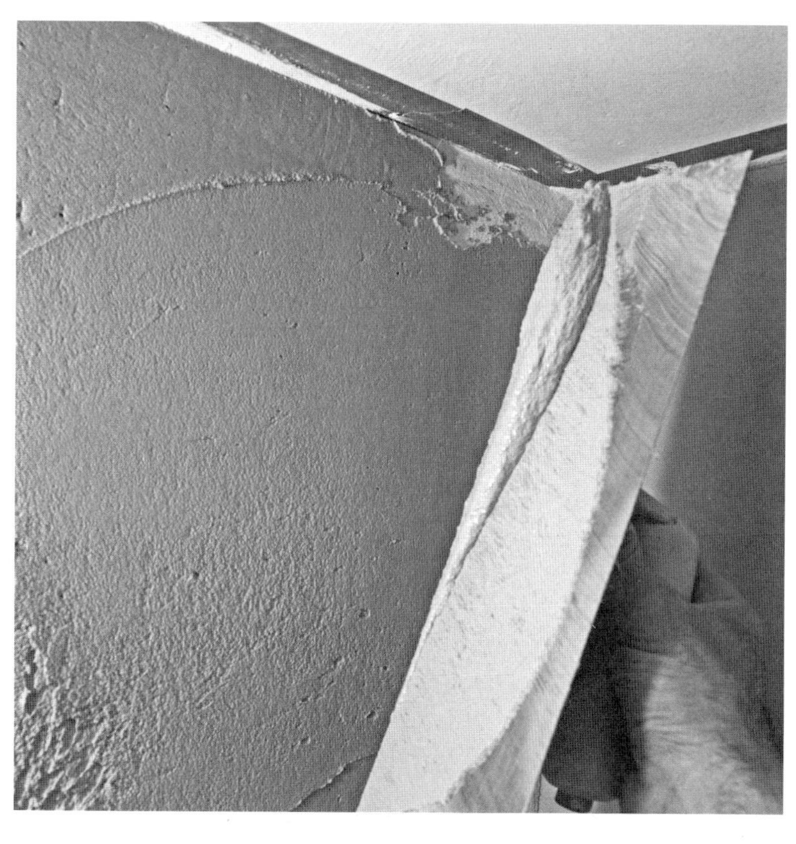

- Over absorbent substrates, hydrate the wall well to manage suction.
- Finish coats are usually applied from the top down to avoid dropping plaster on the finished work.
- Use a hard, rectangular steel trowel to apply mud.
- Angle the trowel 5–20 degrees from the wall and use low (to moderate) pressure. The steeper the angle, the thinner the coat will be.
- Overlap the front (the *toe*) of the trowel a couple of inches over the edge of the last pass to blend them and maintain a consistent depth.
- Maintain a fresh edge. If you're working alone, you'll end up working diagonally down and across the wall so no edge is left for too long.
- After applying mud to a section, several quick passes with a low trowel angle and not much pressure can help flatten the mud.
- Scrape off ridges if needed; if you notice hollows, add a little mud.

Fig. 4.6, 4.7, 4.8: *Finishing steps: 1) With the help of a hawk or a second trowel, get the right amount of mud on your trowel where it's needed to fill holes. 2) Apply with some angle and light pressure to squeeze the mud ahead of your trowel and level the wall. 3) Use long strokes to spread and level the mud.* CREDIT: MICHAEL HENRY

At this point you're trying to level the plaster well. It should look flat and tidy, but not yet perfectly smooth. Try to do all this pretty quickly.

Once the plaster has been applied and leveled, check to see if the mud at the start has firmed up a little. If it has, you're ready to smooth the plaster. Lower your trowel angle, apply very little pressure, and go slow and steady to smooth the plaster. Keep your trowel clean (top and bottom) for finishing.

Use curved strokes to avoid leaving any straight lines showing in your finish.

For uneven walls you may wish to switch to a pool trowel for the smoothing.

Troweling Patterns

For the final finish pass, keep in mind that it's hard not to leave marks at the beginning of a stroke, but it's possible to carefully lift off and leave no trace at the end of your stroke. Start trowel strokes from an edge, or from the rough work, and end them on the more finished part of the wall (or curve back into the rough work).

Curves and corners

The objective with all curves and corners is to get the right amount of plaster in the right place. A carefully leveled corner is a joy to trowel, while trying to do final finish troweling on a corner that isn't ready is a lesson in frustration. A good exercise is to do 95% of all finishing with a hard rectangular trowel, only breaking out the pool trowel or Japanese trowel to erase a few lines and do a final light smoothing.

Inside corners

There are two ways to approach inside corners. Some plasterers prefer to let one wall dry completely before butting up to it. We often prefer to use a corner trowel when both walls are fairly fresh, but you'll need a 90 degree corner trowel designed for cement work (*not* a flared corner trowel designed for drywall taping).

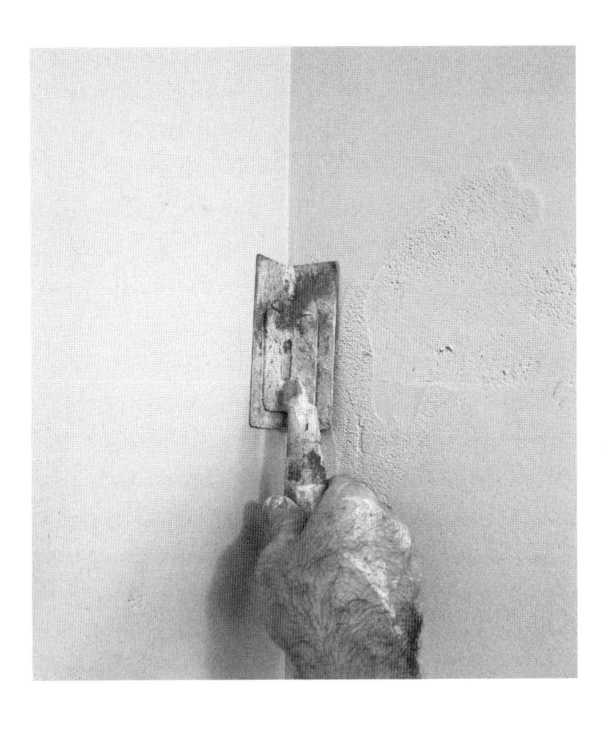

Fig. 4.9: *It takes practice to get the right angle and pressure with a corner trowel, but the resulting corner is crisp.* Credit: Michael Henry

Using an inside corner trowel:

- Level both walls perfectly before using the corner trowel, or the corner won't be straight.
- Try to leave a very small amount of extra mud in the corner, and/or start with some mud on the corner trowel.
- Using gentle firm pressure, try to trowel the whole corner in one stroke.
- Fill any low/rough spots and retrowel the whole corner if needed.
- Run a pool trowel lightly up each wall to erase the lines left by the corner trowel.
- Further touch-ups can be done with a Japanese trowel.

Outside square corners

Outside corners are actually surprisingly easy once you get the hang of them. Here is one way to do outside corners in a few simple steps:

- Level and flatten both walls adjacent to the corner at least a foot or two back from the corner.
- Add mud to the corner until there are no low spots.

- Finish one wall, let any excess mud squeeze to the other side of the corner.
- To get rid of that extra mud, hold the trowel nearly perpendicular to the wall, lightly scrape off the excess mud, leaving a perfectly shaped, slightly roughened corner.
- Smooth the corner with a clean trowel.

Window curves

Window curves are easy if you level them well — to that end, do most of the finishing with a rectangular trowel. Where two curves meet, pay special attention to leveling each perfectly, and the joint almost takes care of itself. A corner trowel paired with a Japanese can be handy for final touches.

Fig. 4.10: *To level a curve, dead-ends on both sides increase your trowel angle and reduce your pressure as you round the curve, so that you start stripping some mud off (rather than pushing extra mud from one side of the curve to the other). By the end of the curve, the trowel may be perpendicular to the wall, leaving a rough surface that can later be smoothed with light trowel strokes.* CREDIT: MICHAEL HENRY

Finishing options

Trowel finish

There's a huge range of possible trowel finishes, from quite smooth, to light trowel marking, to highly textured finishes. The key to a nice trowel finish is usually to do the final finishing pass once the mud has started setting up.

Dry burnish

Dry burnishing is very common with earth plasters, where it can result in a very high-quality finish that still has some subtle texture. Dry burnishing generally happens at the leather-hard stage. A flexible metal or plastic trowel is worked in a rapid figure-eight pattern (or simply back and forth), at a slight angle, to smooth and compress the wall. Continuing to burnish as the plaster dries past leather hard will often polish and add shine to it, which may or may not be desirable.

Wet burnish

Wet burnishing happens around the same time as dry burnishing, but it includes light misting. The result can be an exceptionally smooth, level plaster surface. Generally, the finish left by wet-burnishing is between matte and shiny, and exceptionally smooth. If lime plasters are wet-burnished with black soap (a natural olive-oil-based soap), they can be polished to a shine — particularly if the plaster is past leather hard.

Sponge finish

A sponge finish is achieved by rubbing the entire troweled wall with a slightly damp sponge when the plaster is at or past leather hard. A sponge finish will vary depending on how dry the plaster is and how wet the sponge is — the dryer the plaster and the sponge are, the less sandy and textured the finish will end up looking. For pigmented earth plasters, even if a trowel finish is desired, it's often worth sponging lightly with a barely damp sponge once it is dry, as this will expose aggregate that may add sparkle, and the resulting texture can also make future repairs easier (see Chapter 9).

Consider the Lighting

The lighting in a space will completely change how your plaster looks — the intensity, angle, and color temperature of the natural and artificial light will interact with your plaster, so evaluate your color samples in the space where the plaster will be applied. Compare how it looks at different times of day — direct sunlight is very different from reflected sunlight or different types of artificial lighting.

The *angle* of lighting is often overlooked. If the light will always be fairly flat or head-on, a plaster with more texture might be called for. Very flat plasters can look drab and boring in flat lighting. However, when strong cross-lighting is going to be the norm, such as on a wall that's perpendicular to a south or west window, or one that is lit by a directional light, texture will be greatly accentuated, so a more subtle finish may be preferable — maybe a sponge or burnished finish that reduces imperfections. Strong, angled light will make every imperfection jump out. Work lights can be used to simulate natural lighting while you're plastering — and prevent unpleasant surprises later on.

Troubleshooting

Problems in finish plasters are the same as with any other plaster. Refer to the troubleshooting table for common problems and solutions. The second table lists a few problems specific to veneer finish plasters.

Table 4.1: Troubleshooting Guide

Problem	Likely Causes	Solutions
Plaster poorly bonded, delaminating or falling off	Poorly prepped substrate.	Fill all voids, consolidate loose material before plastering, clean off excessive dust or rubble. Did you use a slip coat, and did you let it set up slightly before plastering?
	Substrate too dry or wet.	If there is a lot of suction, mist/remist substrate and/or slip coat. If there is no suction and a surface sheen of water, it may be too wet for good adhesion; wait before plastering.
	Wall too smooth.	Scratch wall before applying plaster; apply lath or use a bonding/adhesion coat on substrate where necessary.
	Freezing before full cure.	Plaster earlier in season, protect work. If unexpected freeze occurs, try to get some heat on that section of wall if it hasn't frozen too deeply; sometimes you can thaw it gently and save it.
	Incompatible plasters.	Avoid lime over (unstabilized) earth, gypsum over cement, etc. unless an appropriate strategy is implemented.
Weak or crumbly plaster	Not enough binder or poor-quality binder.	Increase binder content; check ratio of clay to silt; add wheat paste or manure.
	Plaster was applied too dry.	Make a wetter mix.
	Poor-quality aggregate.	Use a sharp, well-graded aggregate.
	Excessive additives.	Check/adjust quantities.
Excessive cracking	Needs more aggregate or fiber.	Use test patches to find optimal mix; consider manure as an additive.
	Poor quality aggregate or fiber.	Use a sharp, well-graded aggregate. Make sure sand and fiber are not too fine.
	Applied too thick.	Apply thinner.
	Plaster dried too quickly.	Avoid hot/windy days; protect work.
	Structural movement.	Fill cracks and watch for additional cracking.

Table 4.2: Troubleshooting Guide: Veneer Finish Plasters

Problem	Likely Causes	Solutions
Stains that come through the finish coat	Oil or other stains in substrate, inadequate sealing.	Use a layer of casein to block them, or prime with a paint primer before plastering.
Drywall joints show through	Inadequate sealing, and/or wrong joint compound was used.	Always use a setting type of joint compound, not premix. Use paint primer for adhesion coat over new drywall.
Bubbles of loose plaster	Adhesion problem, often between adhesion coat and substrate.	Make sure surface is free of dust, dirt, and grease before applying adhesion coat. Wait for paint primer to cure fully before plastering. Always use a setting type of joint compound, not premix.

Chapter 5

Earth Plaster Base Coats

About Base Coats

THE BASE COAT PROVIDES the majority of the strength of earth plasters. This strength comes from the depth of the application (typically around ¾ inch) and from the relatively high fiber content (commonly 10–30% of the mix). In general, we can think of base coats as falling into two categories: clay-straw plaster, which might have a ratio of about 1 clay:1 sand:1 fiber; or clay-sand plaster, which might have a ratio of 1 clay:3 sand:<½ fiber. In reality, it's a spectrum, not an either/or situation, but we find that this distinction is a useful way to think about base coats. In general, a more straw-based plaster will have greater strength, especially when applied over a weaker substrate such as straw bale; but a sand-based plaster will be easier to work with and faster to apply for experienced plasterers.

Preparations

Most preparations for earth plasters will be the same as for any other natural plaster (see Chapter 3), but there are some exceptions:

- Use curved corners for most outside corners, because earth plasters are the weakest plaster, and square corners are very prone to damage.
- Stuff and fill voids with slip straw or cob — bare straw stuffing may pull out of the wall during plastering.
- Avoid using metal in the prep, even galvanized metal, because earth plasters tend to cause rust. Metal fasteners will be inevitable in many situations (use coated or galvanized), but large areas of metal lath may rust, eventually causing failure in the plaster.

Table 5.1: Clay-Straw Plaster

Typical ratio	1 clay: 1 sand: 1 straw Clay / Sand / Straw
Application	Hand application, then leveled with a float or trowel.
Advantages	Very strong and resistant to cracking; bridges transitions well. Can be applied by an inexperienced or volunteer crew.
Indicated when:	Substrate is weak, or transitions may cause cracking; or when volunteer labor is available.

Table 5.2: Clay-Sand Plaster

Typical ratio	1 clay: 3 sand: < ½ straw Clay / Sand / Straw
Application	Trowel application, in two passes (see Figure 4.5).
Advantages	Faster to mix and apply.
Indicated when:	Substrate is solid; an experienced crew is available; labor costs are high and/or schedule is tight; additives such as manure are available.

- Because clay is hydrophilic, wood may not always need a barrier from the plaster to protect it from rot.

Mixing and application details

Base coats are usually mixed in a mortar mixer. When a clay slip is used, start with all of the clay slip in place of water, then half the sand, etc. The final consistency of the plaster should usually be sticky and fairly wet, but not sloppy.

If a mortar mixer isn't available, or only a small amount of mix is needed, base-coat plaster can be mixed by hand. It is easiest to add the fiber last, after mixing all the other ingredients with a paddle mixer. The fiber usually needs to be mixed in by hand, or with a hoe, etc.

Earth-base coats need pressure to bond well to the substrate. There are a couple of ways to achieve this. The first is hand application; there's nothing like hands to work plaster into a substrate. After applying roughly the desired thickness of mud by hand, it can be leveled and compressed, usually with a wood float. This is the best (and often, the only) way to apply a very straw-rich base coat, though it is usually slower than trowel application.

The second way to get a good bond to a substrate is to use a rigid steel trowel in two passes, with the first pass pushing mud in, and the second adding depth (a detailed description of the technique is discussed in Chapter 4). When the mud is sufficiently workable, this is the fastest way to apply a base coat. If it is done conscientiously, with the right mix, it's possible to get a bond that is comparable to hand application.

How to process site soil

Maybe you've decided to use site clay, and you have a pile of soil sitting on your jobsite, but in its raw form, it's almost a certainty that it is not ready for immediate use. Maybe it's rock hard, or it might be a mass of incredibly sticky globs; it likely contains stones that need to be sieved out.

There are several good ways to process site clay, but all of them will be easier if the clay sits in a shallow pile, covered by a tarp, through one or more winters. The freeze-thaw cycles help break down the most intractable clumps into a more soil-like product. That said, most people won't have the luxury of that much time.

Processing dry or damp

Soil can be sieved when slightly damp but somewhat crumbly. This requires a good-quality, heavy screen, with about ⅛-inch openings. You will end up pushing the soil through the screen by hand. Even after pushing through a ⅛-inch screen, the result is a little coarser than what you'd get from processing wet, but it is suitable for base coats. If a coarser screen is used, you risk having clay lumps pop out of your plaster.

Soil can also be dried completely by spreading it on a tarp, then pulverized and sieved through a finer screen. This is a good option if you have access to a machine to pulverize the soil, such as a hammer mill. The advantages are that it's easier to store and use on short notice, and it can be screened finer than damp clay.

Processing as slip

Clay slip is just blended clay and water, typically in the ratio of 2 parts clay:1 part water. Site soil is most commonly processed into slip because a very good-quality product can be made with equipment that may already be on site for mixing. Preparation of clay slip should start at least the day before mixing to get a good stockpile. The person in charge of mixing, or someone else who is delegated with making slip, will need to keep a barrel of slip mixing most of the time during plastering. There are two approaches to making slip:

- Soak the clay in barrels for a few days, or weeks, before stirring it. If you have a lot of barrels available, this is a good option. When it's time to make the slip, it will take some effort to get the clay into suspension, but once you do, it will mix easily and thoroughly.
- Add unsoaked clay directly while mixing the slip. To be successful in this technique, it helps to have a very good stirring method, and you should add the clay while maintaining a fairly steady vortex.

Both these methods require similar steps; let's start with the second technique.

- Start with a heavy-duty mixing drill that can withstand running for hours; you also need a barrel with one end open.

- Make a setup to hold the drill in place. Some drills or mixers have attachment points for this; for others, you may need to use All Round steel strapping (found in plumbing section of stores) to fix it — or some other method of your choice. A drill press can be a very elegant option.
- You'll need a very good paddle mixing attachment — our favorite off-the-shelf mixer is the Twister M630. We've also used the Kraft Tools

Fig. 5.1: *On the right is a barrel of clay and water that is kept spinning with a paddle mixer; on the left is an improvised trough to store the processed slip.* Credit: Michael Henry

Fig. 5.2: *One of the best systems we've seen for stirring clay slip is this boat propeller welded to a shaft, used by Quebec plasterer Andrew MacKay.* Credit: Tina Therrien

DC303, a 24" mixer with three small propeller blades to keep the slip moving even as it gets thick. Since the blades are removable from the shaft, a separate metal rod of the correct diameter may be cut to length and used as the shaft.

- Fill the barrel about ⅓ full of water. Measure exactly how much you are using because you need to keep the slip consistent for your recipe.
- Start the mixer spinning in the water before you start adding clay to the barrel — once you have a vortex, start adding the clay. Keep careful track of how much clay you are adding relative to the water. We find about 2 clay:1 water by volume is quite manageable, and that is the minimum required for most recipes (some recipes will require a higher ratio of clay).
- Add all the clay and leave the suspension spinning for at least an hour. Halfway through, stir to the bottom, either by moving the mixer around, or removing the mixer and stirring with a shovel.
- When the slip is well mixed, sieve through a screen (⅛–¼ inch) into a holding tub or trough.
- A slip storage trough may be made using a good tarp or vapor barrier draped over bales. Process at least two barrels of slip before you start mixing plaster — or you'll get behind on slip.
- When you're not using it, keep the slip well covered to keep rain, frogs, leaves, etc. out.
- To process clay that has been slaking in barrels, follow roughly the same strategy, but move the drill and mounting apparatus to whichever barrel is next.

Wedging clay in the mixer (for lime-stabilized earth)

We learned this technique from the *Natural Building Companion,* and for body coats it works well — if the site soil isn't too rocky. This method can only be used for lime-stabilized earth recipes because the lime is necessary to break up the soil structure.

This process is like making cookies — first you cream the sugar into the butter, then you very slowly add liquid, working it in as you go, until the whole thing reaches a nice, even creaminess. In this case though, the butter is the clay, the sugar is dry lime powder, and it all happens in a mortar mixer:

- Shovel or bucket the amount of clay soil that's needed into the mortar mixer.
- Add the lime that's called for in the recipe (commonly about 20–50% of the clay amount) and a small amount of water — just enough that the mixer paddles can start to drag at the clods of clay. The lime will cause the clay to *flocculate* (agglomerate into collections of particles) and become less sticky and plastic. This helps the mixer break it up, and it will slowly begin to cream and soften. A small amount of sand may be added at this stage.
- After 5 minutes, small amounts of mix water can be dribbled in to make the whole thing softer and creamier until it turns into something more like a putty. Add remaining mix water slowly.
- Add the remaining ingredients.

Slip Coats and Adhesion Coats

Earth plaster doesn't adhere well to some common wall surfaces, including straw bales, so an intermediate coat is often needed — usually this is a coat of clay slip, either sprayed onto the wall or applied by hand.

You will likely need a slip coat for:

- Straw bale walls.
- Old plaster walls that are eroded, loose, or dusty.
- Earth walls that are too smooth (may also need mechanical abrasion).

- Any surface where a test patch of plaster can easily be pulled off.

If using bagged clay, add the clay powder to the water, wait a few minutes for it to sink in, then stir it with a paddle mixer.

Clay slip for spraying should be thinner than what would usually be used in plaster. For either site clay or bagged clay, to make a slip for spraying, start with a ratio of about 1.25 clay:1 water, and adjust until the consistency is like thick chocolate milk (but thinner than a milkshake). If you put your hand into the bucket, it should come out wearing a thin, opaque "glove" of clay slip.

Spraying slip coats

- A drywall texture gun hooked up to a medium-duty compressor works very well for spraying slip. An airless electric sprayer with a built-in fan works almost as well, and aerosolizes less clay. From a safety perspective, this makes the airless electric version the better tool.
- On straw bales, keep the sprayer within a foot or less of the wall to force the slip into the bale; apply it heavily enough that you don't see bare straw.
- This is a fairly fast operation; one person can spray one or two walls and then join the trowelers to plaster them.
- Do not mist the wall before spraying slip! If the slip starts to dry out completely before being covered by plaster, you should mist those areas before working them, or you can apply more slip.
- Always wear a respirator to spray slip, and avoid spraying near unprotected co-workers.

The *French Dip*

An alternative to spraying is dipping bales in slip before stacking them. The technique was

Fig. 5.3: *Spraying slip with an airless electric texture sprayer.* Credit: Deirdre McGahern

invented by French builder Tom Rijven. It was dubbed the *French Dip* at a California straw bale building conference in 1998. Note that only the two exposed sides of the bale are dipped in slip. The Dip results in an excellent plastering surface, with great penetration into the bale and immediate weatherproofing. However, bales become much heavier (roughly double in weight), which can be a workplace safety issue. Two people are usually needed to place bales. It's also very messy:

- Fill an actual — or improvised — bathtub with clay slip.
- If bales are very shaggy, pre-trim the sides to be dipped/plastered.
- Submerge one side of a bale in the slip, push down gently and evenly for roughly 15 seconds to force clay into the bale.
- Remove the bale and scrape some of the excess slip back into the tub.
- Dip the other side. It is a two-person job to remove it from the tub safely.
- Lay the bales out to dry partially, to soft but not sticky (2 hours or more). Depending on the weather, the bales could be placed later the same day, or the next day.
- With the *French Dip*, it takes longer to build a wall than with bales that haven't been dipped, given that you have to dip the bale, and then let it dry somewhat. A team of six people should

be able to lay in between 30–50 bales a day, provided that they have made the equivalent of three bathtubs of slip ahead of time.

Compression, Scouring, and Scratching

Earth plaster base coats will usually be either scoured or scratched. See Chapter 4 for a description of each. Scratching needs to happen relatively soon after application (often within a few hours), while the plaster is still workable. Scouring and compression are usually done after the plaster has hardened somewhat — to the leather-hard stage.

Table 5.3

Use compression and scouring when:	Finish coat will be ⅛" or less. Using straw-rich base coats. A very flat wall is important.
Use a scratch trowel or comb when:	Finish coat will be >⅛". Using sand-rich base coats. Finish coat will be lime.

Cracks: Are They a Problem?

Fine cracks in the base coat are generally not considered a problem. Larger cracks may come through the finish coat, or they may cause weak points in the plaster — they should be filled and checked to ensure that the plaster on either side of the crack has no movement. If the plaster is loose or moves, it may need to be stripped off and the area replastered. This is a rare occurrence, however, and it may indicate a problem with the substrate or prep.

Most plasterers aim to avoid having large cracks in their base coats. French builder Tom Rijven welcomes them. "Oh, les belles fissures!" he writes ("oh, the beautiful cracks!"). He says that with his body coat, you need to worry if it *doesn't* crack. This is part of his system, and he knows what to expect. We don't really understand it, but it has helped us relax if we have a few cracks in the base coat (we still aim to avoid them). Rijven fills the cracks before applying a finish coat.

Drying and Aftercare

Base coats should be allowed to dry completely before applying the finish coat. This can take several days under very good drying conditions, but it might take up to two weeks — or more — in cool, damp weather. Scheduling may make it tempting to apply the finish coat when parts of the base coat are not 100% dry. Don't do it. Slow and uneven drying is likely to cause patchy discoloration and mold growth on the finish coat. You want the base coat completely dry because you will need to mist it before applying a finish coat, and you want to be able to rehydrate the wall evenly.

Modifying Recipes and Evaluating Tests

Most problems with plaster can and should be solved by adjusting the ratio of clay:sand:fiber. This should always be your starting point. If that doesn't solve the problem, there may be an issue with one of the ingredients — poor-quality sand or too much silt mixed with the clay, for example. Or there could be problems with the substrate; maybe it was too dry when the plaster was applied, or it was too loose or dusty. Minor issues can be remedied with additives such as manure or wheat paste, but you should look at additives as a way to improve a plaster that is already pretty good, not to fix a plaster that has real problems. See the end of Chapter 4 for a troubleshooting guide for plaster base coats.

Recipes

Plasterers tend to experiment and collect recipes over the years. Favorite recipes get tweaked, with a pinch more of this and that, and are shared with fellow natural builders, just as family recipes are passed on through generations. The intent of this book was to create a "cookbook" for plasterers — something we haven't yet seen. We invited natural builders from around the world to share recipes for the book, with the idea that having a selection of recipes to choose from is a really good starting point for novice or experienced plasterers. Many friends and colleagues accepted our invitation, for which we are eternally grateful. The recipes in this book each make one "batch" of plaster, which usually equates to a 6 ft³ mortar mixer-full, using the equivalent of one bag of binder.

Recipe

Project Karyne Base Coat from Site Soil

A simple, easily workable, trowel-on base coat using clay-rich site soil.

Contributed by Camel's Back Construction

Introduction

We've used this simple base coat as an interior plaster on buildings where we had access to a very clay-rich soil (trucked in from a local source). The hardest part of the process is making the clay slip, which needs to be reasonably thick for this recipe to work, so you'll need to make some slip ahead of time. Otherwise, it is fast to mix and apply. If wheat paste is used (vs. dry starch), it tends to feel a bit too wet, but it pushes into the substrate well and still sets up hard. Depending on how wet the mix is, it can be difficult to build up more than about ¾ inch with this recipe, but you can add extra straw or cut hemp sliver to fill low spots. We typically apply this in a fairly thin application (⅝ inch), then use a fiber-rich finish coat, applied about ¼-inch thick to add strength. We first used this recipe on a slip-straw greenhouse that was built as part of an environmental project to commemorate an 8-year-old girl named Karyne, who died from cancer.

Recommended for:	Depth	Advantages	Limitations
Slip-straw buildings, well-prepped straw bale buildings where limited filling is needed, or other solid natural walls.	⅝–¾" (16–19 mm)	Uses site soil. Fast to apply.	Time to make slip and wheat paste. Depth limit.

Ratio (by Volume)	Quantity
1 clay slip	25 L
2 sand (up to 2.5 if less straw is used)	50 L
0.15 wheat paste (or .04 pre-gelatinized starch)	3.75 liter (or 1 L starch)
1 chopped or mulched straw	25 L
½ cup of borax per mixer load (optional)	(½ cup)

The Details

Clay — A fairly thick slip is necessary for this recipe (at least 2 clay:1 water). The plaster may feel somewhat wet and sloppy, but as long as it doesn't slump, you're ok. A little more straw will help dry it out slightly, and will allow you to build thickness better, but if you find yourself dramatically increasing the straw, your slip may be too thin and you might not have enough clay in your mix (directions for making clay slip were given earlier in this chapter).

Sand — Masonry sand, or any well-graded sand will do (cover with a tarp in case of rain).

Wheat paste — We like the added strength of starch, and in a base coat we like dry starch because it's less likely to promote mold growth if the plaster dries slowly. However, dry starch can be hard to source, and we've mostly used this recipe with wheat paste and when drying conditions were good. When using dry starch, you may need to add extra water.

Straw — May be chopped to 2–4 inches, but preferably mulched a little finer than this. If it's just chopped, the plaster will be less workable, and you might need to cut the straw back slightly. If mulched, the plaster will be easier to apply. You'll be able to increase the straw, which will help dry out the mix. Mix different lengths of straw for the strongest plaster.

Borax — Optionally, you can add ½ to 1 cup of borax to a mix to reduce surface mold inside buildings, where drying time is slow. With good ventilation in summer, this shouldn't be necessary. We rarely use borax.

Application and finishing

Apply clay slip adhesion coat and let it set up for 5–10 minutes before plastering. To apply the plaster, use a trowel to push a thin bonding layer into the substrate. You should use a lot of pressure and a steep angle. Then add depth and level the coat with a square steel trowel or a wet wood float, aiming for an average depth of around ⅝–¾ inch. Scratch with a comb or scratch trowel.

Mixing

Mix in a mortar mixer.

Mixing Sequence	Notes
clay slip	
½ the sand	
starch	Dry starch may be premixed with the sand.
straw	Adjust to suit.
remaining sand	
water, if needed	Unlikely if wheat paste is used.

Coverage

Approximate, assuming an average application depth of ⅝–¾":

Unit of material	Coverage	Calculation (Area in ft^2)	For 1 ft^2
1 yard clay	960 ft^2 (89.2 m^2)	wall area/960 = ___ yards clay	.001 yards (.8 L)
1 yard sand	480 ft^2 (44.6 m^2)	wall area/480 = ___ yards sand	.0021 yards (1.6 L)
1 bale straw OR 1 large bag chopped straw	300 ft^2 (27.9 m^2) 150 ft^2 (13.9 m^2)	wall area/300 = ___ bales straw wall area/150 = ___ bales straw	.0033 bales .0067 bags (.8 L)
1 qt (L) paste OR 1 lb (0.45 kg) dry starch	8.3 ft^2 (.77 m^2) 37.7 ft^2 (3.5 m^2)	wall area/8.3 = ___ L paste wall area/37.7 = ___ lb starch	0.12 qt (L) .027 lb (.012 kg)

Recipe

Easily Workable Base Coat Using Bagged Clay

A fast trowel-on base coat.

Contributed by Camel's Back Construction

Introduction

This is an easy recipe to apply — it mixes fast and goes on fast. The main holdups we've had are not having enough wheat paste ready for the full day of mixing, and having to strain lumps out of manure. We developed this recipe early on in our switch to earth plasters. It's descended from a recipe that was shared with us by potter Linda Taylor. Although we like using clay dug up on site, we often used bagged clay when site soil is poor — but the choice of bagged clay is important (see Chapter 2).

This plaster won't handle fills much deeper than 1 inch. On very uneven walls, pre-fill deep holes with a straw-enriched mix, or choose a different plaster. This mix was developed for straw bale walls, but could be used on any natural wall system.

The Details

Clay — Hawthorn Bond is a pottery clay with very high plasticity (comparable to some ball clays), and it has good diversity of particle sizes. Ball clays could probably be used in this recipe, but test patches should be done first to determine if extra sand is needed.

Sand — We use masonry sand. Any sharp sand with a diversity of particles up to ⅛ or 3⁄16 inch should work well.

Straw — We commonly use a chainsaw to cut the straw for this recipe. If the straw were mulched, more could be added to the recipe, and depth tolerance and workability would probably improve.

Manure — We have used horse and cow, actively blended with mix water and strained through ¼-inch screen to remove lumps.

Recommended for:	Depth	Advantages	Limitations
Well-prepped straw bale buildings, or other natural walls.	⅝–¾" (16–19 mm)	Fast to mix and apply. A good choice when clay soil is hard to source but bagged pottery clay is available.	Time to make wheat paste, sourcing manure and clay. Depth limit.

Ingredients

Ratio (by Volume)	Quantity
1 Hawthorn Bond pottery clay	1 bag (20 L)
2.5 sand	50 L
0.1 wheat paste (or .025 pre-gelatinized starch)	2 L (500 mL starch)
0.5 + chopped straw (2–4+ inch cut)	10–15 L
0.25 manure (if omitted, increase wheat paste to 2.5 L)	5 L
0.75 water (may vary significantly depending on moisture in sand)	15.5 L

Mixing

Mix in a mortar mixer, let it turn for 5–10 minutes. Workability tends to improve if the mix is allowed to sit for a few hours before application, but it isn't a necessity. Extra mix may be left covered overnight.

Mixing Sequence	Notes
¾ water	
½ the sand	
wheat paste	
manure + ¼ water	Mixed into a slurry and sieved.
bag of clay	
straw	Add slowly.
remaining sand	
water if needed	

Application and finishing details

Most walls systems, especially straw bale, will need to have clay slip sprayed on shortly before the base coat is applied. Apply plaster at least 5–10 minutes after spraying slip. Apply with a hard square steel trowel, pushing a thin pass into the substrate with a lot of pressure before adding more mud and building up depth. We usually scratch with a scratch trowel or comb when it is slightly firmed up, and then we use a ¼-inch finish coat. For thinner finish coats, you can rub the base coat with a sponge, or you can scour it with a wood float when it is approaching leather hard.

Coverage

Assuming an application depth of ⅝–¾":

Unit of material	Coverage	Calculation (Area in ft²)	For 1 ft²
1 bag clay (20 L)	26 ft² (2.4 m²)	wall area/26 = ___ bags clay	.038 bags (0.76 L)
1 yard sand (765 L)	400 ft² (37.2 m²)	wall area/400 = ___ yards sand	.0025 yards (1.9 L)
1 bale straw OR 1 large bag chopped straw	630 ft² (58.5 m²) 315 ft² (29.3m²)	wall area/630 = ___ bales straw wall area/315 = ___ bags straw	.0016 bales .0032 bags (0.38 L)
1 quart (L) paste OR 1 lb (0.45 kg) dry starch	16 ft² (1.49 m²) 73 ft² (6.8 m²)	wall area/16 = ___ qt (L) paste wall area/73 = ___ lb starch	.063 quart (L) .013 lb (.006 kg)

Recipe

Straw-Clay Mud

A hand-applied heavy sculpting base coat for any natural wall.

Contributed by Carole Crews (with credit to Bill Steen)

Recommended for:	Depth	Advantages	Limitations
This mix is good on any type of wall, but especially on uneven walls. Smooth walls can be prepped with clay and glue-soaked burlap, or a glue-sand adhesion coat.	1+" (25+ mm)	Handles depth and holds shape well; can be sculpted or used to flatten uneven walls.	Slower to mix and apply than some other plasters.

Introduction

Straw-clay mud is sticky, malleable, and maintains any shape you give it — when mixed to the right consistency with enough long straw. It's useful for flattening walls or for adding sculpted elements, including niches or shelves. This plaster will add tremendous strength to a wall and help with leveling uneven surfaces. The biggest downside for production-oriented plasterers will be the need to hand-apply the mix to the whole wall. Others may really enjoy this process, especially if a mud-party is organized. At an early colloquium at the Black Range Lodge, Bill Steen and Carole Crews were talking and realized they mixed mud the same way — so credit for this recipe is shared between them, as well as hundreds or thousands of traditional builders scattered across the Southwest.

Ingredients

Ratio (by Volume)	Quantity
1 water, adjusted to suit	4 gallons
2 high-clay-content loam	7–9 gallons
1.5–2 loose long straw	6–8 gallons
0 to 1 sand	0–4 gallons
0.25 manure (optional)	1 gallon
.06 starch paste	1 quart

The Details

Soil — The soil only needs to be sieved if it has rocks that need to be removed, otherwise use it as-is. Let the soil soak in the water for 5–10 minutes before mixing.

Straw — The straw should be used long, it does not need to be chopped, but it should be loose and broken up. This can be accomplished by opening a bale in a chicken pen and letting the chickens pull it apart, or by spreading the straw on the floor of the worksite where it will be walked over. The straw can also be pulled apart by hand as you add it to the mud.

Sand — The quantity varies according to the sand content of the soil. If the soil is mostly clay with little sand, you may want to add 1 part sand to it; however, this recipe usually will work with no sand added. Masonry sand works well if sand is desired.

Manure and wheat paste — They aren't needed for interior plaster (although manure will help reduce cracking), but they are recommended for exterior plasters. Because of slow drying times, wheat paste should be avoided inside because it will promote mold and may get stinky. The starch paste may be wheat or rice paste.

Mixing

If you have access to a backhoe, you can use it for mixing — the soil may be lifted and dropped as water is sprayed on and straw is slowly added. If the soil is sifted, it may be mixed in a mortar mixer. Spin soil and water and slowly add the straw, let it mix for about 10 minutes. Smaller quantities can be mixed by hand in a wheelbarrow. The mix should hold together without being stiff, but it shouldn't be shiny-wet. If it's too wet, add a bit more soil.

Mixing Sequence	Notes
water	
clay soil	Soak 5–10 minutes before mixing.
wheat paste	Omit for interior plaster.
manure	Optional for interior plaster.
sand	Depending on soil.
straw	Add gradually.

Application details

Over straw bales, apply a bonding coat of the adobe dirt (clay loam) and water, just thick enough to pick up with your rubber-gloved hand. Work it into the bales by hand. Alternately, clay slip may be sprayed onto the bales. Before this is completely dry, but once it has dried to the leather-hard stage, it's time to apply the straw-clay mud plaster. If the bonding coat is too dry, mist it before applying this coat. Apply the thick layer of straw-clay mud by hand, taking large amounts and slapping them against the wall with some force to create a good bond, then smooth it out with your palm or a wet wooden float. Hold the float with two hands and use your body weight to push against the plaster. You can put numerous blobs on the wall, making sure to press their edges together well to get a good bond, then flatten them all at once.

Finishing techniques

Don't use a metal trowel, which will promote large cracks on this thick body plaster; the wood float should leave enough texture that scratching isn't necessary for thin finish coats. You can achieve a harder, smoother surface by passing over it again with a wet wood float when it is leather hard. At this stage, you can also add detail by pressing or cutting into the plaster.

Coverage

This is approximate coverage, and should be used as a guideline only:

Unit of material	Coverage	Calculation (Area in ft^2)	For 1 ft^2
1 yard soil (765 L)	300 ft^2 (28 m^2)	wall area/300 = ___ yards soil	.003 yards (2.3 L)
1 yard sand (765 L)	600+ ft^2 (56+ m^2)	variable	variable
1 bale straw	100 ft^2 (9 m^2)	wall area/100 = ___ bales straw	.01 bale
1 quart (L) wheat paste	14 ft^2 (1.3 m^2)	wall area/14 = ___ qt (L) paste	.07 quart (L)

Recipe

Lime-Stabilized Base Coat Using Bagged Clay or Site Clay

A reliable, strong, trowel-on base coat.

Contributed by New Frameworks

Introduction

This is a very versatile and dependable recipe for a base coat over straw bale and other natural walls. It is easy to mix, and easy to apply. It sticks well and is easily smoothed and built out for desired thickness, up to 2 inches in some cases. Average depth for this coat is 1 inch. Scratches well and dries firm for next coat. The lime adds a plastic workability to the mix, and it also modifies the manure odor to an earthy, mineral smell that makes this recipe a pleasure to work with. We have found the lime content in the clay base coat to be necessary for providing a chemical relationship to the subsequent lime finish coat, which is typical for our applications, especially on the exterior where saturation and freeze-thaw conditions make lime and clay compatibility a higher risk. A quality mechanical bond (i.e. good *tooth* provided between the base and finish coat) is also critical to address this issue. The lime content also serves to "stabilize" or strengthen site soils that have questionable or "at-the-edge-of-acceptable" clay contents. The lime provides for faster drying and raises the alkalinity of the coat (thus reducing the potential for surface mold); this is a benefit in our moist region of the northeastern United States — and to many construction schedules. If site clay is used, this recipe will need to be tested and modified to adjust for other unknown subsoil content beyond the clay content.

Recommended for:	Depth	Advantages	Limitations
Well-prepared straw bale buildings, or other natural walls.	¾–1½" (19–38 mm)	Fast to mix and apply, very durable, can receive pure lime finish. A good choice when clay soil is hard to source, but bagged pottery clay is available. Site clay can also be used. Ideal is 30% clay content in subsoil.	Sourcing, gathering, and working with quality manure. Lime content requires skin care and protection. Insufficient lime can result in weakened mix.

Ingredients

Ratio (by Volume)	Quantity
1 ball clay (bagged) OR site soil of 30% clay content can be used (with testing)	1 bag = 6.1 gallons (23 L)
3.5–4 sand	21–24 gallons (80–92 L)
0.5 lime (bagged, powdered Type S hydrated lime)	3 gallons (11.5 L)
1 chopped straw (cut to 2–4+ inch)	6 gallons (23 L)
0.5 manure (cow or horse — optional, but increases strength and reduces cracking)	3 gallon (11.5 L)
1–1.5 water (may vary significantly depending on moisture in sand and manure)	6–9 gallon (23–35 L)

The Details

Clay — Ball clay provides good strength and stability as a binder, chosen here over the more delicate kaolin clays or more expansive clays that can lead to cracking.

Sand — A clean, sharp sand with varied particle size is desired. We have had luck with granite sand and ¼-inch minus masonry sand. Avoid sand with a high silt content, as it will promote cracking and make the plaster weak.

Lime — Bagged, dry, powdered Type S hydrated lime can be found at most masonry supply yards. Much of the lime in North America has magnesium carbonate in addition to calcium carbonate, so the firing process requires a process called *autoclaving,* which makes our lime different from the traditional $Ca(OH)_2$ limes found in much of Europe. The upshot of this is that we can use our powdered lime directly, without soaking it (as was thought necessary for many years in the North American natural building field).

Straw — We use a weed whacker in a 33-gallon trash can to cut the straw for this recipe, or a hammermill. Pre-shredded straw is also available in many farm and garden stores.

Manure — We have used horse and cow. Both should be fresh and free of bedding or other material. We used to slurry the manure and then add that to the mix, but have since found that adding manure without additional water into the mixer with a bucket of sand and lime in the beginning of the mix breaks it up nicely and helps it integrate into the rest of the mix. We have found that spending the time to source and collect fresh, quality manure is worth it for easy incorporation into the mix. Cow manure features more micro-fibers, which creates a stickier mix as the enzymes react with the lime and clay, and helps improve the set of the plaster. Horse manure features more macro-fibers, as the horse does not digest the cellulosic fiber in its fodder.

Mixing Sequence	Notes
all the sand	
all the clay	
all the lime	Mix all three dry.
½ the water	Pay attention, and add more water if it binds.
manure	Watch until it is broken up and integrated.
the rest of the water	
straw	
water, if needed	We send a test trowel-full over to the wall to test it and see if the sand or water content needs to be adjusted on the first batch.

Mixing

Mix in a mortar mixer. We use a vertical shaft, 12 ft^3 mortar mixer. If using a horizontal shaft mortar mixer, use smaller quantities because the equipment won't mix larger volumes as easily.

Application and finishing details

Most walls systems, especially straw bale, will need to have clay slip sprayed on with a drywall texture gun and air compressor shortly before applying the base coat. Apply plaster at least 5–10 minutes after spraying slip, or pre-wet with water just before plastering. Apply with a hard square steel trowel, pushing a thin pass into the substrate with a lot of pressure, and working into the straw in multiple directions before adding more mud and building up depth. We call this *one coat, two passes,* meaning you apply a very well-adhered thin layer in your first pass, then build up thickness to shape the wall and smooth it with the second pass before moving onto a new area. Scratch well with a scratch trowel or comb. A low angle to the scratch tool and medium-firm pressure works best to create grooves without tearing the plaster.

With special finishes, a plastered wall becomes shiny — reflecting light in a room. Plastered walls have a unique look and feel that invites touch. Credit: Athena Steen

Artistic sculptures brighten and adorn this lime plaster home in Killaloe, Ontario. Plaster can sometimes invite spontaneous design that may not have been part of the original plan. Credit above and left: Michael Henry

Above: The Steen's paint shed at Canelo is reminiscent of an old English potting shed with its order, variety of tools, and color. Credit: Athena Steen

Left: An intricately carved mandala provides a most wondrous piece of art on this plastered wall. Credit: Athena Steen

While green is one of the more expensive pigments, it readily blends into a wooded surrounding in this pigmented coat of silicate dispersion paint.
CREDIT: TINA THERRIEN

Bottom left: Interior bale walls allow for arched doorways, carved-in bookshelves, and niches. Soft shades of a painted plaster wall showcase antiques in this straw bale home outside of Peterborough, Ontario.
CREDIT: TINA THERRIEN

Bottom right: Trellised vines help shade this owner-built load-bearing bale home from the sun in the summer months, yet allow for full passive solar gain in the winter. CREDIT: TINA THERRIEN

This tree was sculpted on a whim by the client and author Michael Henry. It's built over an armature of metal lath stuffed with straw, and shaped with a thick coat of earth plaster. The plaster was Camel's Back Construction's pigmented finish plaster with lots of fine hemp sliver added (chapter 6). Credit: Michael Henry

Bottom left: An experienced plasterer is able to obtain these sleek curves around window and door openings.
Credit: Athena Steen

Bottom right: When used as a decorative element, plaster adds beauty and interest to a wall. Credit: Deirdre McGahern

Top : A tadelakt wall pigmented with yellow ochre is beautiful and very water resistant, but the technique is challenging and labor intensive. Credit: Michael Henry

Middle right: Tiles and plaster marry well in this bathroom. The plaster is Camel's Back Construction's stuc/faux tadelakt recipe (chapter 7). It is splash resistant, but not waterproof. Credit: Michael Henry

Bottom left: A custom tadelakt sink, though tricky to plaster, is a funky and unique addition to this bathroom. Homeowners need to treat the sink carefully and clean it with black soap solution. Credit: Michael Henry

Bottom right: This bale wall has been finished with a coat of tadelakt plaster. Applying tadelakt around a curved window opening is a challenging task, not for the faint of heart. Tadelakt may affect the permeability of exterior bale walls, so be cautious when choosing to use it. Credit: Michael Henry

A lovely shade of green in this tadelakt shower surround is achieved with chromium oxide deep green pigment. Troweled, stoned, soaped, and waxed, it is now highly water resistant and holds up well against daily showers.

Credit: Michael Henry

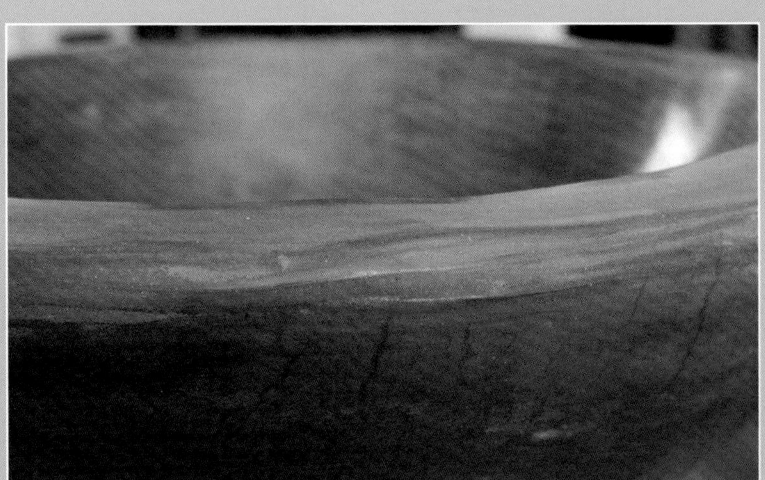

A tadelakt vessel sink, pigmented with Mars violet. The core of the sink is formed first with a hydraulic lime or cob mixture; the slope around the drain needs to be shaped carefully to prevent pooling of water. The backsplash is colored with a Mars violet pigment from a different manufacturer.

Credit: Michael Henry

These photos were taken at the Builders Without Borders Ecohouse that was built in Washington, DC, in 2008, in workshops.

Top left: A variety of trowel sizes make up every plasterer's arsenal of tools. Bill and Athena Steen trowel on clay plaster. Credit: Catherine Wanek

Top right: A clay alis is first painted onto a plastered wall and then buffed with a sponge. Credit: Catherine Wanek

Middle left: Athena Steen at work, detailing some decorative carving in an earthen plaster. Credit: Bill Steen

Middle right: Kaki Hunter paints a decorative trim around the doorway of this lime-plastered building. Credit: Catherine Wanek

Bottom left: A plastered niche is a perfect spot for a sculpture, or a candle (not left unattended, of course). Niches add beauty to a plastered wall, capturing and reflecting light in interesting ways. Credit: Catherine Wanek

Bottom right: The finishing touches. Credit: Catherine Wanek

A delightful cobwood garden wall built in part by workshop participants. The cob is raised on a foundation, and clay tiles protect the wall from precipitation. Many hands make light of this work. Credit: Catherine Wanek

A pigmented exterior earth plaster works on this house because of the extensive overhang, doubling as a porch roof. Functional and beautiful . Design and plaster by Chris Magwood and Jen Feigin. Credit: Michael Henry

Athena Steen's flair for creating masterpieces that incorporate design, texture, and color is admired by plasterers around the world.
Credit: Athena Steen

Recipe

Straworks' Baseball Diamond Mix

A versatile trowel-on lime-stabilized base coat using infield mix.

Contributed by Straworks Inc.

Introduction

Recommended for:	Depth	Advantages	Limitations
Well-prepped straw bale walls where a lime finish coat will be applied.	⅝–1" (16–25 mm)	Fast to mix and apply and can be troweled on thick or thin. A good choice when clay/sand mixes made for baseball diamonds are available.	Experiments will be needed to adjust recipe to mix available in your area.

Where there are baseball diamonds, tennis courts, and BMX tracks, there may also be commercial clay/sand mixes that are suitable for plastering! It turns out that the ratio that works for baseball infields works for us, too. We use Hutcheson Granite Infield Mix (it's 25% clay to 75% crushed granite) as the basis for our base coat plaster in our clay/lime plaster system. It comes in a dump truck, so there are no bags to deal with, and, because it's supplied to baseball diamonds across southern Ontario, ordering is easy and the supply is stable. Hutcheson Sand & Mixes is an aggregate producer located in Huntsville, Ontario. A comparable company in your area may be able to supply you with something similar or make you a custom mix.

We add lime, finely chopped straw, loose straw, and water to make a base coat plaster that's troweled over wet clay slip on straw bale walls inside and out. Once dry — which is, at most, a week in the summer — we finish it with hydrated lime plaster.

We like this mix because it is locally sourced and versatile: it can be troweled on thick or thin in order to flatten an undulating straw bale wall without cracking issues. The mass of chopped straw allows it to be applied thick to build up low spots, and the plasticity of the clay/lime binders allows it to spread thinly in a high spots. The value of being able to do both with one plaster is huge. The other advantage is that it can be left to dry completely and then easily be hydrated to an even dampness prior to applying the finish coat of lime plaster. Mist the base coat the night before, the morning of, and as needed during plastering. We find this is an advantage over straight lime plaster base coats, which, once cured, can be difficult to hydrate evenly.

Ingredients

Ratio (by Volume)	Quantity
1 water (may vary significantly depending on moisture in sand)	8.5 L
0.5 lime putty	4.25 L
1.5 chopped straw (¾–1 inch)	12.75 L
0.5 loose straw from bales (6–8 inch)	4.25 L
4 granite infield mix from Hutcheson Sand & Mixes	34 L

The Details

Lime putty — One 50 lb bag of lime:14 liters of water mixed in a bucket with a paddle mixer. Makes about 24 liters of thick lime putty.

Chopped straw — We use a wood chipper with a ¾" screen to make a fine chopped straw.

Loose straw — Straw sweepings off the floor after installing bales in the wall.

Granite Infield Mix — 25% clay:75% crushed granite. Particle sizes range from ⅛ inch to powder.

Mixing

This is for one mix in a 6 ft³ mortar mixer:

Mixing Sequence	Notes/Quantities
water	17 L
lime putty	8.5 L
chopped straw	25.5 L
loose straw	8.5 L
Granite Infield Mix	68 L

Application and finishing details

Spray clay slip directly onto well-prepared straw bale wall. It can be sprayed in advance of the trowelers so long as it will not dry before base coat is applied. Sprayers and those nearby should wear a respirator to avoid inhaling vaporized clay. Apply base coat plaster with hard square steel trowels over wet clay slip. Build up low spots with more base coat and push it thinly over high spots to achieve a more or less flat wall. Scratch base coat with a scratch trowel to ⅛ to ³⁄₁₆ inch max.

Coverage

Assuming an average application depth of ¾":

Unit of material	Coverage	Calculation (Area in ft²)	For 1 ft²
1 bag of hydrated lime	150 ft² (13.9 m²)	wall area/150 = ___ bags of lime	.0067 bags (0.15 kg)
1 gallon lime putty	24 ft² (2.2 m²)	wall area/24 = ___ gallons of lime	.042 gallon (0.16 L)
1 bale straw	388 ft² (36 m²)	wall area/388 = ___ bales straw	.0026 bales
OR 1 contractor's bag chopped straw (120 L)	255 ft² (23.7 m²)	wall area/255= ___ bags chopped straw	.0039 bags (0.47 L)
OR 1 contractor's bag loose straw (120 L)	800 ft² (74.3 m²)	wall area/800= ___ bags loose straw	.0013 bags (0.15 L)
1 yard (765 L) Granite Infield Mix	607 ft² (56.4 m²)	wall area/607 = ___ yards of mix	.0016 yards (1.26 L)

Recipe

La Couche de Corps

A body coat with fermentation.

Contributed by Tom Rijven

Introduction

French plasterer Tom Rijven's mixes are distinguished by their use of fermentation liquids. This is also the only recipe we know of that includes sawdust. The quirks come with some definite benefits. This body coat is typically applied at a depth of 1 inch, but it can be built up as deep as 6 inches. Because of the fermentation that occurs in the plaster, it's more water resistant than most earth plasters. These benefits come with the trade-off of requiring greater time and labor investments.

Recommended for:	Depth	Advantages	Limitations
Straw bale or other natural walls.	1+" (25+ mm)	Handles depth well, can be used to flatten uneven walls. Some water resistance.	Slower to mix and apply than some other plasters. Time needed for fermentation, and labor for crack filling.

The Details

Clay — The clay slip is made by dry screening stones from the clay, then slaking the clay in water directly in the mixing tub. A complicated slip-making setup is not recommended.

Fiber — The fiber is usually a mix of straw and a greener, fermentable fiber such as hay, grass clippings, ferns, etc. Generally about 50:50 reinforcing fiber to fermentable fiber.

Sand — The sand is well graded up to ⅛ inch (3 mm) maximum size. Masonry sand will work fine. The sand is mostly in here to act as a mediator between the other ingredients, and to keep the straw from floating out of the mix.

Sawdust — Sawdust or shavings can be obtained from a local sawmill.

Fermenting liquid — Ah, the mysterious fermenting liquid! This can be made by taking ensilaged corn or any other sugar-rich material, covering it with water and fermenting it, covered in a black container in the sun for a few days. A sauerkraut smell indicates fermentation.

Ingredients

Ratio (by Volume)	Quantity
3 thick clay slip	7.9 gallons (30 L)
3 fiber	7.9 gallons (30 L)
½ sand	1.3 gallons (5 L)
1 sawdust	2.64 gallons (10 L)
½ fermenting liquid	1.3 gallons (5 L)

Mixing

Mix with a paddle mixer in a 60 L tub.

Mixing Sequence	Notes
slaked clay	Soak overnight.
fermenting liquid	
fiber	Add the fermenting liquid to the clay, then the fiber.
sand	
sawdust	Finally, add sand and sawdust. Always start with the most liquid substance at the bottom of the container, and mix well.

Test the cohesion of the recipe by throwing handfuls of the paste inside the mixing container. It is ready when you can form a layer about 6 inches (15 cm) thick. After mixing, the plaster is covered well with black plastic and fermented for several days before using it.

Application details

On straw bale walls, this plaster is usually applied over bales that have been dipped in clay slip during construction (the "French Dip"). This clay coating is rehydrated before plastering and a coat of thick clay slip is applied over it with a scratch trowel. The plaster is applied before this slip coat has dried. The body coat is applied by hand, starting at the top and working down, pushing mud firmly against the wall and upward to join to the mud that has already been applied.

Finishing techniques

Once mud has been applied and roughly leveled by hand, run over it with a leveling board about six times to flatten it, scraping off the high points and filling the low spots. It is then smoothed with a pool trowel. Allow it to set for about an hour and then scratch it. After it has dried, scrape off any high points, and fill larger cracks. Gouge the cracks open and fill them with a mix of 1 earth to 2 sand. Now you're ready for finish plaster.

Recipe

Super Sticky Upside-Down Mix

A sticky clay and straw combination.

Contributed by James Henderson, Henderson Clayworks

Introduction

This plaster is the true do-it-all problem solver. It is so sticky and lightweight that it can be used to quickly shape anything, anywhere. I use it to rough in the upside-down sections of window reveals on straw bale houses. It works great as a big gap filler for those strange problem spots. It is also the go-to mix for roughing in artistic reliefs. The combination of good quality clay slip and short straw (with no fines) is the key.

Recommended for:	Depth	Advantages	Limitations
Upside down work, repair work, relief work.	¼–3" (6–76 mm)	Lightweight and super sticky. Will not crack. Good substrate for top coat.	Hard to get a smooth finish.

The Details

Clay — Any really sticky clay with little to no sand. Bagged clays work great. Site clay works best if first screened through a ¹⁄₁₆" (1.6 mm) screen.

Straw — Chop the straw however you can. Screen through a ½" screen to get down to 2" minus straw. Then screen though a ¼" screen, but keep the stuff on top of the screen. You don't want the fines. Keep them for other plasters.

Application and finishing details

Did I mention this stuff is sticky? It is best to apply with a bare hand. Keep a bucket of water handy to keep your hands clean. Shape and smooth with a wooden float. The float will have to be cleaned very often in the bucket of water to keep it from sticking to the plaster. After a few hours, the plaster can be compressed with the wood float or a small steel tool.

Ingredients

Ratio (by Volume)	Quantity
1 Hawthorn Bond or Lincoln fire clay or good site clay	1 bag (20 L)
1½–2 chopped straw 1–2" (screened to drop out fines)	30 to 40 L
water, just enough to make a thick shake consistency	approx. 15 L

Mixing

Mixing Sequence	Notes
½ water	
clay	
½ water	Mix with a drill and mixing paddle.
straw	Mix the straw in with your hands; keep adding straw and kneading until it is well combined. This is a fluffy mix, not watery.

Chapter 6

Earth Plaster Finish Coats

About Finish Coats

EARTH PLASTER FINISH COATS are mostly used inside buildings because their poor weather resistance makes them unsuitable to exteriors. Their advantages include beauty, ease of repair, and low embodied energy. The main disadvantage is their relative softness, so ease of repair is counterbalanced by the need for more frequent repair.

Earth plaster finish coats can be broadly divided into two categories: *thick* and *thin*. Thick finish coats are applied up to ¼ inch and have fiber in them. This type of plaster is called for when the base coat is rough or uneven, which is often the case for straw bale walls or any other wall that requires a lot of leveling or has a particularly soft substrate. A fairly fiber-rich finish coat has some structural role in the plaster system.

Thin finish coats are typically applied at ¹⁄₁₆ inch or less, and may have very fine fiber or none at all. They play more of an aesthetic role and less of a structural one, and they may be applied over natural walls that are well leveled and don't need to be built up structurally. All finish coats share the requirement to be tough and resistant to the hard realities of daily life. One of the ways to toughen up earth plaster is to add a lot of starch paste or a similar binder. This is optional if walls are going to be painted with a durable paint, but if an earth plaster is to be left unsealed, a starch, casein, or similar binding agent is generally required.

How Ingredients Change the Properties of Finish Plasters

Depth tolerance

The two main variables you can adjust to control how thick your plaster can be applied without cracking are *fiber* and *sand*. Adding more of either one proportional to the clay will allow you to build depth. Adding coarser fiber or sand may also allow for a thicker plaster coat.

Texture

This is an important aesthetic quality. Generally, more clay, finer sand, and less fiber make for a smoother plaster. But sometimes a rough texture is what people want; a finish with more variation and some fiber showing can be very beautiful. The first thing to figure out is what you're going for, and then adjust the recipe and application techniques accordingly.

Hardness

The most common way to get a harder plaster is to add starch, basically gluing the plaster together. Adding starch can affect workability, texture, and reparability. Compressing plaster after it has partly dried will also improve hardness and strength.

Color

This is dependent on the color of the clay and sand, the texture (more texture appears darker), whether you add pigment, and of course the lighting on the wall being plastered.

Repairability

The most repairable plaster will be a simple clay-sand plaster, but additives are often necessary to get the hardness and strength that is needed. In any case, finishing techniques usually affect repairability more than ingredients do.

Preparations

Mixing and application details

Earth finish plasters are usually applied quite thin, so relatively little plaster is needed to cover a given wall area. Because of this, they are often mixed in a tub with a paddle mixer. Thick coats with plenty of fiber will still commonly be mixed in a mortar mixer.

Application of earth plasters will vary by recipe, but will generally be similar to any other finish plaster — level the wall very carefully, aiming for a very even coat of mud before trying to make it pretty.

Fig. 6.1: *Earth finish coats are commonly mixed with a paddle mixer, but for heavy applications a mortar mixer may be preferable.* CREDIT: MICHAEL HENRY

Finishing

Finishing earth plasters can be confusing because so many variations in choice of tools and timing are possible, even when working with the same plaster. We'll try to demystify it. The first thing to understand about earth plasters is the stages they go through as they dry:

Wet

When earth plasters are first applied, they may be very wet and hard to work with. At this stage, you're mostly aiming to get mud on and get the wall level; don't worry about getting out fine trowel lines, but do take the time to remove highs and fill lows — it shouldn't look messy. Apply fairly large areas of wall using a hard rectangular trowel; at the same time, keep an eye on the work behind you, which still needs additional finishing.

Stiff

If the substrate is very dry (unsealed earth or masonry), the plaster will stiffen up almost immediately; if not, it might take 10–30 minutes, maybe more. This stage, where the plaster is still quite soft but stiff enough to work easily, may be an opportunity for you to do most of your finishing with one or two quick passes of a rectangular or pool trowel.

Sticky

After it becomes stiff, earth plaster usually becomes sticky, grabby, and hard to trowel. It is best to leave it alone for a while at this point.

Leather hard

Leather hard is what it sounds like — it's hard to leave a fingerprint in the plaster, but easy to dent it with a fingernail. This might happen one or more hours from the time of application, depending on drying conditions. At this point, it can be burnished thoroughly with either a pool trowel or a Japanese trowel. Light troweling with a Japanese trowel can leave drag marks at this stage, so burnish with enough pressure and motion to compress the wall. Small trowel marks can still be burnished out at this point.

Hard burnish

After the leather-hard stage, the wall can still be compressed and smoothed for a while, but it is too hard to remove most marks. This is the perfect time to really polish the wall, if that's what you want, but a plastic trowel must be used because steel trowels will leave black marks. You can also sponge the wall at this stage.

Dry

Once the wall is completely dry, you can use a slightly damp sponge to bring out sparkle if there's marble dust in your plaster. Marks can't be removed except by rehydrating the wall, and then either troweling or sponging them out. One of the easiest ways to get an even sponge finish is to let the wall dry completely, rehydrate it with a good sprayer, wait a few minutes for the water to penetrate, and sponge with a slightly damp sponge. This avoids different parts of the wall being at different stages of drying (which results in different textures when sponged).

Unsealed Plasters

There are few finishes more beautiful than an unpainted, pigmented plaster. If you're already plastering, it can also save the time and expense of painting, and it's one of the least toxic wall finishes available. However it's not for everyone; without a protective coating of paint, the plaster is particularly vulnerable to marks and damage, but also very repairable.

When you're at the planning stage of a project, you should consider the strengths and weaknesses of different plasters. For example, we avoid unsealed earth plasters in kitchens. And we avoid putting any natural plaster on a corner where it's likely to get bumped a lot — next to a door threshold, for example. Trim it out with wood if that's an option, or at least make sure exposed corners are rounded.

Plan for a finish you (or the homeowner) will be able to maintain — most people can learn to repair a troweled plaster with some dedication, but if a homeowner wants to be able to easily repair their own plaster, a sponge finish may be the way to go.

When you are installing an earth plaster, always save some mix for future repairs. The easiest way to do this is by drying lumps of plaster and storing them in a container someplace away from moisture. It's also a good idea to write the pigment ratio on the container — in case of larger repairs or to match for another wall, etc.

Recipe

All-Purpose Finish Plaster

A versatile finish plaster that can be used on uneven walls.

Contributed by Camel's Back Construction

Recommended for:	Depth	Advantages	Limitations
Interior straw bale or other slightly undulating walls that are protected from the elements.	3⁄16" (5 mm)	A strong plaster that can be applied with some depth.	Sourcing manure. Plaster tends to be sticky initially.

Introduction

We use this earth plaster as an interior finish coat on straw bale walls that have a sand-rich base coat that may not be perfectly leveled. Because we can apply this plaster slightly thicker (about 3⁄16 inch — or more with extra fiber), this lets us apply the base coat a little less perfectly and make up the difference on the finish coat. At ¼ inch or more, cracking starts to become a problem, but adding extra-fine fiber or finely chopped straw can fix this. The high wheat paste content (and the optional manure) result in a very strong plaster. It's usually painted with a silicate dispersion, milk, clay, or other natural paint; it could be covered with an alis or pigmented earth plaster. It contains enough wheat paste that it could be left as-is if a low silica clay is used. This finish coat is used on the interior, usually paired with a lime-earth system on the exterior. The above-average amount of wheat paste in this recipe makes it initially quite sticky, which takes some getting used to, but it is a very workable plaster.

Ingredients

Covers approx. 150 square feet

Ratio (by Volume)	Quantity
2 clay	28 L
6 masonry sand	84 L
1 wheat paste	14 L
0.1 hemp sliver	1.4 L
0.5 manure (optional)	7.5 L
0.5 finely chopped straw (optional)	7.5 L

Recommended substrates/prep

This versatile plaster can be used as an interior finish coat for any natural wall system; it is usually applied over an earth plaster base coat. It is particularly valuable for straw bale because it can help slightly with leveling the base coat. The base coat should usually be scratched to maximize adhesion, though it will stick well to a floated base coat if a thin application is used.

The Details

Clay — We've mostly used this recipe with bagged clay, and sometimes a kaolin clay such as Tile 6. When greater strength or depth of application are desired, we'll use Hawthorn Bond. However, site clay can be used if it is sieved well to remove stones and lumps of clay. Generally, it needs to first be processed into a fairly thick clay slip, and then sieved through a ⅛-inch screen.

Sand — Masonry sand is usually the best option, but any sand with good particle distribution will work (the largest particles should definitely be smaller than ⅛-inch).

Wheat paste — Making wheat paste can slow down the entire crew if you're not careful. You should make your wheat paste ahead of time — but not more than one day ahead at most, or it will spoil. Substituting pre-gelatinized wheat starch will speed up the mixing process and cut the likelihood of bad smells or mold on the surface of the plaster. The dry powder can be added at about ¼ the volume of the wheat paste to get a similar amount of glue. This recipe can be used with a little less wheat paste; as low as about 8% will still result in a very strong — and sticky — plaster.

Fiber — Any very fine fiber such as flax, fine hemp, or synthetic poly fiber can be used. Hemp cut to ½ inch is ideal for this plaster, but it may not be available to you. Hemp can be added up to ½ part or more, but it starts to affect finish texture and workability around ¼ part or more. We've used poly fiber because it's easy to come by from masonry supply stores, but we only turn to poly when we need smaller quantities of fiber in our plaster.

Manure — Manure is optional, but helpful when applying a thicker coat over uneven walls. Horse or cow manure can be used, but either one should be sieved through a ⅛-inch screen. This can be done dry, or it can be made into a slurry with some of the mix water before sieving.

Finely chopped straw — We don't usually use it, but chopped straw may be used instead of, or in addition to, fine fiber when the finish coat will be applied fairly thick. It must be quite fine, such as straw that has been screened through a 1-inch screen. Even so, it will result in a coarser finish texture.

Mixing

This plaster is usually mixed in a mortar mixer, but for a small crew and thinner applications, mixing with a paddle mixer is an option. The wheat paste and manure (if made into a slurry) will contribute the majority of the water that is needed, so be sure to add these before adding much water to the mix. While adding fiber, take extra time to break it up as much as possible. If using hemp, it may also be best to add it in alternating layers with other ingredients, to prevent it clumping together.

Application details

Allow the mix to sit for several hours before applying it. It can also be left overnight if well covered. This brief aging period will reduce the stickiness and greatly improve workability. This is particularly important if using bagged clay, which takes time to fully hydrate.

Fig. 6.2: *The base and finish coats of this plaster were made from bagged clay (the finish is Camel's Back all-purpose finish plaster — recipe given in text).*

Credit: Michael Henry

This plaster can be applied between ⅛ and ¼ inch in depth; if a lot of extra fiber is added, it may be built up even further. We usually apply it with a hard square trowel except on very uneven walls, where we prefer using a sturdy pool trowel.

Finishing techniques

This coat can be applied and mostly smoothed almost immediately, or within about half an hour, with a pool trowel. After that, it becomes very sticky for a while. A quick burnish pass can be made a few hours later to smooth out trowel marks. Burnishing is mostly an aesthetic choice — we don't worry about compressing this plaster because the wheat paste gives it good strength, but the results achieved by burnishing a few hours later are very good. A sponge finish may lift some of the fiber above the surface depending on the type of fiber used; particularly when synthetic fibers are used, a sponge finish should be avoided.

Coverage

For coverage of 3⁄16":

Unit of material	Coverage	Calculation (Area in ft²)	For 1 ft²
1 bag Hawthorn Bond	109 ft² (10.1 m²)	Wall area/109 = ___ bags clay	.0092 bag (0.183 L)
OR			
1 bag kaolin clay	153 ft² (14.2 m²)	Wall area/153 = ___ bags clay	.0065 bag
OR			
1 yard site clay	4,600 ft² (427.3 m²)	Wall area/4,600 = ___ yards clay	.00022 yard
1 50 lb bag sand (14 L)	30 ft² (2.8 m²)	Wall area/30 = ___ bags sand	.033 bag
OR			
1 yard sand (765 L)	1,400 ft² (130 m²)	Wall area/1,400 = ___ yards sand	.00072 yard (.55 L)
1 qt (L) wheat paste	12 ft² (1.1 m²)	Wall area/12 = ___ qt (L) wheat paste	.083 qt (L)
OR			
1 lb starch (0.45 kg)	54 ft² (5 m²)	Wall area/54 = ___ lb starch	.019 lb (.0083 kg)
1 oz fine fiber (min.)	150 ft² (13.9 m²)	Wall area/150 = ___ oz fiber min.	.0067 oz (0.19 g)
1 qt (L) manure (optional)	24 ft² (2.2 m²)	Wall area/24 = ___ qt (L) manure	.042 qt (L)

Recipe

Pigmented Finish Plaster with Fiber

A forgiving, intermediate-textured, pigmented earth plaster.

Contributed by Camel's Back Construction

Introduction

This is a recipe for a pigmented plaster that can be used over most walls, including drywall. It is usually applied about 1⁄16 inch over drywall, but may be applied 1⁄8-inch thick or more without excessive cracking, depending on the quality of your sand and how much fiber you add. The idea is to show off the beauty of the earth, so it is pigmented instead of relying on a covering of paint. Therefore, you need to use bagged kaolin pottery clay in this recipe because of its whiteness and its low silica content (if there is any dusting, toxicity will be very low). A moderately high wheat paste content binds the surface so that dusting will be minimal, and the wall is more resistant to damage. This is a good base recipe that can easily be modified with fiber additions. This recipe makes a good sculpting mix that can be applied up to an inch thick in places and troweled smooth if you increase the sand slightly (2:6 clay:sand), increase the hemp to 1 part (10% of the mix), and use ¼ part pre-gelatinized starch in place of the wheat paste.

Recommended substrates/prep

Can be applied over most fairly flat surfaces; it is commonly used over prepared drywall. It will also work well on natural walls with well-prepped earth plaster — usually as the final coat in a three-coat system. To use it as the finish in a two-coat system, increase the hemp (up to 1% is workable) and prepare the base coat well. The base coat may be left smooth, or (preferably) slightly roughened with a sponge or scoured with a wood float. Mist the wall ahead of your work on absorbent walls (not on drywall).

Recommended for:	Depth	Advantages	Limitations
Veneer coat over drywall, or a finish plaster on natural walls where there's a good base, and a pigmented finish is desired.	1⁄32–1⁄8" (1–3 mm)	The same plaster can be used on a variety of types of walls. Beautiful.	Less durable than many paint finishes; requires occasional repair.

Ingredients

For coverage of 250 ft² at a depth of 1⁄16":

Ratio (by Volume)	Quantity (for a depth of 1⁄16")
2 kaolin clay	16 L
5 fine sand	40 L
0.50 wheat paste (or 0.83 dry starch and adjust water)	3–4 L
.02 hemp sliver (½" cut) — or more as needed	0.1 L+
1.75 water	14 L

The Details

Clay — Use only kaolin clays for whiteness and low silica content. Our favorite for good plasticity is Tile 6 (or 6 Tile).

Sand — Use a fine sand (preferably 30-mesh or finer), a white marble sand, silica sand, or sieved masonry sand. To get deeper colors (especially reds), use all or part sieved masonry sand. For brighter colors, use only white sand. If your sand has poor particle size diversity, you'll need to use a finer mesh size and apply it thinner, or blend sands to get a better mix.

Wheat paste — the amount given is the optimal amount in most cases. It can be as much as doubled to make a harder, more adhesive plaster; however, workability will be reduced.

Fiber — The best fiber is fine hemp sliver, cut to ½ inch. If you can't get a good fine hemp sliver, textile flax might be the closest analogous fiber. Fiber must be teased apart well or it will clump.

Mixing

It is usually mixed in a 50–60 L tub with a paddle mixer. Start with water, add the kaolin clay and let it sink into the water. Give it a quick stir with the mixer; this will make a slip. Add all of the remaining ingredients. If using dry starch, mix it with the sand before adding it to avoid lumps (important), also break up the hemp fiber by hand and layer it with the sand so that it mixes in thoroughly. The pigment is usually mixed with enough water to make a thin slurry, then added part way through mixing.

Application details

Apply thin, with a square trowel, using a fairly steep trowel angle, then lightly scrape off excess where necessary. If you can hear the trowel scraping the substrate, you have gone too thin. Smooth and level well with square trowel, then, before moving on to the next wall section, pass over it with a large, good-quality stainless steel pool trowel. It should look pretty good at this point.

Finishing techniques

For a fairly smooth texture, pass over the wall again with the steel trowel when it is at the leather-hard stage. This will also help compress the wall, creating a harder surface. To leave more texture on the wall, pass over the wall with the steel trowel after it has set up slightly but before it has reached the leather-hard stage; this will leave some drag marks on the wall that can be pleasing to the eye (this technique is tricky, and probably not for beginners); another pass at the leather-hard stage is still advisable.

Experiment with texture at home. A pass over the wall with a sponge the same or the following day can be done in lieu of, or in addition to, the pass with the pool trowel. Burnishing with a plastic Japanese trowel when it is at, or slightly past, the leather-hard stage will leave a very fine finish, but the whole wall will need to be burnished to create a consistent finish. An attractive finish can also be achieved by sponging over the wall with a thin slurry of the plaster in tight swirls, after the wall has dried. The end result is variations in color that can resemble clouds; it's a bit messy and time consuming, but eminently repairable. This recipe won't polish to very smooth or shiny unless whiting or other fines are added in place of some of the sand.

Coverage

For an application of 1⁄16":

Unit of material	Coverage	Calculation	For 1 ft^2
1 L sand	6.25 ft^2 (.6 m^2)	Wall area/6.25 = ___ L sand	.16 L
1 bag clay OR 1 L clay	500 ft^2 (46.4 m^2) 16 ft^2 (1.5 m^2)	Wall area/500 = ___ bags clay Wall area/16 = ___ L clay	.002 bags .063 L
1 L wheat paste OR 1 L starch	50 ft^2 (4.6 m^2) 360 ft^2 (33.4 m^2)	Wall area/50 = ___ L paste Wall area/360 = ___ L starch	.02 L .003 L
100 mL hemp fiber	260 ft^2 (24.2 m^2)	Wall area/2.6 = ___ mL fiber	0.4 mL

Recipe

Silty Subsoil Dolomite Sand Top Coat

Bagged sharp dolomite sand helps overcome dustiness in a silty finish coat.

Contributed by James Henderson, Henderson Clayworks

Introduction

Silty subsoil is a tricky beast. Unlike purer clays, a silty clay will both crack and be dusty in a plaster mix. This is due to the amount of microscopic sand (or silt) mixed in with the clay. I have yet to find someone who can tell me how to remove it — and oh, how I've tried.

The best solution to the problem is to mix the subsoil into a thick clay slip, add ½ part fresh horse or cow manure, and let the whole lot ferment for a week or a month. However, the time demands of a plaster job often do not allow for the luxury of a week or more fermentation. Many people add wheat paste in this situation. As I am not a fan of wheat paste, I started playing around with sands — sharp sands, dolomite sand. I have found that a good, sharp sand blend and lots of straw can produce a fine top coat plaster — *without* fermentation or wheat paste.

Will it work for all subsoils? I doubt it.

Recommended for:	Depth	Advantages	Limitations
Top coat, over clay plaster.	⅛–¼" (3–6 mm)	Allows the use of silty subsoils in top-coat plaster, without relying on fermentation or manure.	Cost and availability of dolomite sand.

The Details

Clay — I use the subsoil from my property. I also buy subsoils from quarries that produce gravel in my area. They have to wash the clay off the gravel before it can be used in concrete, so it is a waste product for them.

Sand — Dolomite sand. Dolosand is the brand I use.

Silica sand — Lane Mountain is the brand I use. You may not need the silica sand — it all depends on the subsoil and the finish you want. Experiment.

Straw — I use wheat, barley, or rice straw. I have a shredder that cuts it up. I like to have lots of fines in the straw for this mix, as well as pieces up to 1½ inches in length. Rice straw is definitely the best choice for this plaster due to its rough texture and grippy surface.

Ingredients

Ratio (by Volume)	Quantity
1 part clay-rich, but silty subsoil, screened through a ⅛-inch (3 mm) screen	20 L
1 part #70 dolomite sand	20 L
1 part #30 dolomite sand	20 L
¼ part #120 silica sand	5 L
1 part finely chopped straw–1½ inch minus	20 L

Mixing

Best mixed in a pan mixer. (All plasters are best mixed in a pan mixer.) Add the water, add the clay, a

Mixing Sequence	Notes
20 L water	
20 L clay	Mix with a paddle on a drill or in a pan mixer.
20 L #30 dolosand	Add and mix.
20 L straw	Add and mix.
20 L #70 dolosand	Add and mix, may need a little more water now.
5 L #120 silica sand	Add and mix.

little sand, the straw, then the rest of the sand. Let it mix for 5 minutes before using.

Application and finishing details

I prefer to apply this with a hawk and a jigane trowel (a stiff Japanese trowel, excellent for applying a base coat). After a little while, compression can be done with the jigane or a wood float. When leather hard, smooth out the trowel marks with a flexible plastic trowel. A sponge or a misting bottle can be used to help get a good finish.

Recipe

Fat Plaster

A thick, weather-resistant, durable, sculptable earthen plaster for interior or exterior application.

Contributed by Kaki Hunter and Donald Kiffmeyer

Introduction

We had been hired by the US Government to build a Ranger Station for the Bureau of Land Management which challenged us to develop a new standard for a durable, mold-resistant, earthen plaster. Hence, Fat Plaster was born! Fat Plaster is just that. It's a very thick coat of earth plaster that goes on in two stages, the second immediately following the first. Fat Plaster is super strong, weather resistant, and fun to apply. If the color of the earth is attractive, the plaster can be troweled smooth and left as the wall finish, or it can be lightly textured to receive subsequent coats of a more refined plaster, alis (clay paint), or milk paint. Fat Plaster reduces the majority of prep time by allowing the use of coarser materials. This means less time spent screening! Whoopie!

Recommended for:	Depth	Advantages	Limitations
Straw bale, Earthbag, light clay straw, adobe, and even Sheetrock walls!	Up to 2" thick	One application. Weather and mold resistant. Strong, durable, looks great!	I can't think of any, well, maybe your endurance.

Ingredients

Ratio (by Volume)	Quantity
3–4 clean water	3–4 gallons (11.3–15 L)
1.5–2 chopped straw	2 #10 cans (6–7 quarts)
1.5–2 dried grass clippings	2 #10 cans (6–7 quarts)
4 paper cellulose	4 #10 cans (12–14 quarts)
6 clay-rich soil, screened to ¼" or ½" (0.625 or 1.25 cm)	6 shovels
18 sand (screened to ¼" or ½" [0.625 or 1.25 cm])	18 shovels
1⁄12 part borax dissolved in twice as much cool water	2 cups dissolved in 1 quart of water
	NB: #10 cans are the large coffee cans, about one gallon or a shovelful.

The Details

What makes Fat Plaster FAT?

Just as earth-building soil or earthen plaster benefits from well-graded sand, we have found the same benefit to apply to the fiber in a base-coat plaster. Although many options for fiber exist, we found three varieties that suit our purposes the best: chopped straw from chaff up to 1½ inches (3.75 cm) long; sun-bleached grass clippings up to 1½ inches long; and recycled shredded paper or cellulose (like the kind used as blown-in insulation).

Straw provides bulk and adds tensile strength throughout the matrix. The fine grass clippings also add tensile strength, while making the plaster malleable and easy to sculpt. The shredded paper is the crème-de-la-crème, making the plaster super creamy while providing tensile strength on a micro level. The combination of three distinct sizes of well-graded fiber mixed with clay-rich soil of (approximately) 25–30% clay and 75–70% well-graded sandy soil has produced a strong, crack-resistant plaster that is fun and easy to apply.

This mix will make an average 4 ft^3 wheelbarrow load. Everyone tends to develop his or her own style of mixing. Usually, it's best to start with a portion of the water (say ⅔ of the total amount). We add the diluted borax in the beginning to thoroughly amalgamate it with the water. We add the clay next, to ensure it is completely saturated. Then we add the sandy soil and fiber, using a little more or less of each to achieve the desired consistency. We strive for a consistency that is firm enough to mold into a ball, yet pliable enough to sculpt with a trowel. Let it be said that Fat Plaster is clay and fiber-rich, sticky yet firm, plastic and pleasurable. It should not sag or feel sloppy in your hand or on the wall.

Mixing Sequence	Notes
½–⅔ water	Start with ½ to ⅔ of the total water in your mix.
diluted borax	Mix until well amalgamated (if using flour paste, add at this stage).
clay-rich soil	At least 25–50% clay.
fiber and sand	About equal parts fiber and sand.
feel	This is more about feel than percentages! Keep adding all as needed.

Fig. 6.3: *Applying Fat Plaster paddies over smear coat on a light clay straw wall.* CREDIT: OKOKOK PRODUCTIONS

Application and finishing details

Stage one: The Smear Coat

The first stage is the mortar coat, or *smear coat*. The smear coat is made from finer ⅛-inch screened clay and sandy soil (same as the Fat Plaster, but without the straw or grass clippings) mixed to a consistency of rich sour cream. The only fiber we add to the smear is the shredded paper. The smear coat should feel very sticky, yet thick enough that when you scoop it into your hand, it won't slip through your fingers. Apply the smear coat by working it into the straw bales or spreading it evenly over your prepared surface (not too thick, not too thin).

Stage two: Fat Plaster Paddies

Form a glob of Fat Plaster into a 1-inch-thick paddy about the size of your whole hand and slap it over the moist smear coat. The paddies will suck onto the wall like glue. Apply as many as you can keep up while the smear remains moist. If the smear dries out, the paddies won't stick! We like to slap on at least a 3 × 3 square foot (or larger) area of Fat Plaster and then trowel it evenly with a wood float, adding little bits of plaster to any low spots. After the Fat Plaster has cured to a leather-hard surface (still green, but firm), it can be hard-troweled smooth or left rough for additional finer plaster coats, lime, or clay paints.

For interiors, we use the Fat Plaster recipe with a percentage of borax to inhibit mold and add extra strength (about 2 cups per 4 cubic foot wheelbarrow load). The borax adds strength and water resistance and makes the plaster more compatible with lime plaster.

Coverage

In general, one 4 ft^3 wheelbarrow of Fat Plaster spread 1-inch thick covers about 8 square feet.

Troubleshooting

If cracking is observed:

- Reduce the amount of water, OR
- Increase the amount of sand, OR
- Increase the amount of fiber, OR
- Try all three!

Fig. 6.4: *A freshly Fat Plastered wall with sculpted arch detail over windows.* CREDIT: OKOKOK PRODUCTIONS

From Athena Steen

We use a plaster similar to Fat Plaster over bales, and any other substrate that needs a lot of fill, bevels, sculpting, etc. I work mostly with a wood float, except for the final finishing. If I am turning it into a finish, which often I am, I compress it as it is drying to seal up the pores and "smash" the straw or other long fiber into the clay to prevent cracking. I can get super sharp edges as it compacts. Before it dries completely, I add a bit of clay slip to help fill the pores and hit it with a semi-stiff metal trowel. I enjoy the blend of the smooth, filled-in surface with the long exposed fibers.

Fig. 6.5: *Athena uses a sculpting plaster mix to create divine art on the wall.* CREDIT: BILL STEEN

Recipe

Finish Coat with "Mayonnaise"

A durable, somewhat weather-resistant finish coat.

Contributed by Tom Rijven

Recommended for:	Depth	Advantages	Limitations
Finish coat over clay plaster.	3⁄16" (5 mm)	Creates a hard finish coat; adds some water repellency.	Fromage blanc may be hard to procure in North America.

Introduction

This recipe is a fairly durable, somewhat water-resistant finish. This finish trowels on nicely. The fromage blanc is used as a glue to make the mixture creamier and more pleasant to use.

Ingredients

Quantity
10 L thick clay slip
20–30 L well-graded sand
250 g (9 oz) fromage blanc
"mayonnaise" — about 0.5 L

Recommended substrates/prep

This finish coat can go over any base coat of earth plaster or clay slip, which could be over adobe blocks, concrete, plaster board, or many other surfaces. The substrate must be rough and able to absorb water to be suitable for an earth plaster.

The Details

Clay — Where to source it:

- Check geologic maps for soil types
- Building sites
- Roadworks
- Stores — powdered or solid clay can be purchased
- Quarries

We sieve our clay slip through a 3⁄16" screen (5 mm) for finish plasters (easier to sieve liquid clay than dry clay).

Sand — By sifting the sand, and using sand with differing grain sizes (from 1⁄16–1⁄8" [1.5–2.9 mm]), we find this increases the hardness of the finish coat and creates good cohesion.

Fromage blanc —This is a soft cheese originating from France and Belgium, which is hard to find in North America. It's similar to Quark, which is also hard to find. The closest readily available dairy product might be Greek yogurt, or perhaps an unripened goat cheese. However, it is easy to make your own fromage blanc, and only takes about a day. You'll need a starter culture, which can easily be ordered online. Tom uses a low-fat fromage blanc.

"Mayonnaise" — Mayonnaise is a combination of egg yolks and linseed oil. For a volume of 30 L of finish plaster, you will need .5 L of "mayonnaise." Whilst whisking the yolks of 5 eggs, slowly add enough linseed oil to create .5 L. Later, the liquid white of the eggs is added to the finish plaster as a hardener.

Mixing

In a 60 L basin, pour in the most liquid ingredient — the clay slip (10 L bucket). Add 250 g fromage blanc, and then the appropriate amount of sand — decided after testing (i.e. 3 buckets [30 L]). 40 L of mixture creates only 30 L of plaster (the clay "disappears" between the sand grains). Mix everything with a drill and mixing paddle, and when it is well mixed, add .05 L of "mayonnaise" to the 30 L of finished plaster. Add water if necessary, noting down the amount for the next batch. The desired consistency is like fromage blanc (like a thick sour cream).

Application and finishing details

This finish coat trowels on beautifully. Apply the plaster with a float (or trowel) at a 45° angle, working upward. During the upward movement, tip the float slightly forward, closing the angle so that at the end of the movement it is almost parallel to the wall. The thickness of the coat is achieved by the amount of pressure applied. Apply in a thin coat, keeping a wet edge as you work and working each new stroke into the previous stroke. Keep the float clean between each application by scraping it quickly after each pass on the edge of the hawk. Start at the top of the wall and work down. For a smoother finish, take two large sweeping passes with a long flat blade (like a drywall trowel), leave it to rest for 2–3 hours, and then use a pool trowel for the finishing touch. You can play with the appearance of the finish by changing the angle of the trowel.

Recipe

Finish Coat Using Bagged Clay

A very workable trowel-on finish coat.

Contributed by Endeavour Centre

Recommended for:	Depth	Advantages	Limitations
Any prepared substrate, including natural plasters, drywall (raw or painted), masonry, or plywood.	⅛–¼" (3–6.5 mm)	Fast to mix and apply. A single-coat system for finishing a wide range of substrates.	Depth limit. Better for flat, smooth substrates. Not for use in very wet areas.

Introduction

This is a well-tested recipe that leaves room for experimentation and variation that can give lots of different appearances. We find that this mix is just as fast to apply as painting a room — since it only requires one coat — but is much richer and more beautiful than paint. The materials typically cost less than paint, too.

The exciting thing about this recipe is that it can go on almost any substrate, included painted drywall. This plaster can bring a touch of natural building into any room or home! Walls can be prepared with a mixture of wheat paste and sand that is painted or rolled onto the surface.

We've used this plaster in lots of workshops and find that first-time plasterers are able to create really attractive finishes. It's an empowering plaster!

Ingredients

Ratio (by Volume)
4–5 dry bagged pottery clay (Hawthorn, ball, kaolin, etc.)
10 sand (finely screened)
0.75–1 wheat paste
1 whiting (calcium carbonate, marble dust, etc.)
0.25–0.5 pigment
water (may vary significantly depending on moisture in ingredients)

The main caveat with this plaster is that it goes on thinly and doesn't like variations in thickness.

The Details

Clay — Hawthorn Bond is a pottery clay with very high plasticity and a good diversity of particle sizes. We've also used ball and kaolin clays in this recipe. It's always prudent to make test patches to fine tune the recipe.

Sand — We use masonry sand and sift it through a bacon splatter screen (which is a fraction of the cost of sieve screens). The sand has to be quite dry to make it through the fine screen. The particle sizes and distribution changes the plaster noticeably. More fines will give a smoother surface; more large particles give the surface a bit of texture. There is no right or wrong, but it's worth experimenting a bit to get the right result.

Whiting — Most masonry sand does not have enough very fine particles, so we add calcium carbonate (talc) or marble dust to increase the fines. You can bump up or turn down the amount of whiting for different results.

Flour paste — The amount of flour paste changes the hardness of the finished plaster. This quantity can be varied quite a bit from the proportion in the recipe to make the plaster harder or softer. Our flour paste is 1 flour, 2 cold water, 4 hot water.

Pigment — We've used a wide variety of natural and synthetic pigments in this recipe, and the colors tend to remain quite true to the pigment color. If you are using a lot of pigment, reduce the sand and/or whiting quantity by the same amount, as the pigment is playing the same role in the mix.

Application and finishing details

Walls are prepared with a mix of wheat paste and sand. Be sure that the sand in the preparation is of the same particle size as will be used in the plaster. The paste/sand mix can be applied with a brush or roller. Try to get an even coating, and be sure to stir the mix frequently, as the sand will settle to the bottom. If you leave lumps on the wall surface, they will show through the plaster. The paste/sand takes about 2–4 hours to dry before it's ready for the plaster.

We have found that the paste/sand mix is beneficial, even on substrates that may not require it (such as natural plaster base coats). The wheat paste creates a moisture barrier between the finish plaster and the substrate, which keeps the plaster workable for a longer time than if applied directly on the substrate.

We use a small, stiff, straight-edged trowel for application. This mix doesn't like to be overworked, so apply roughly over a large area and then finish it with as few trowel strokes as possible. Keep the trowel edge clean between strokes.

After the initial application (15 minutes to 1 hour, depending on drying conditions), we usually go back and compact the surface. This takes down any unwanted trowel marks and compresses the surface, tightening the pores and making the plaster harder. The degree to which you do this is up to you; sometimes we don't do it at all; sometimes we do it vigorously. The final appearance and texture changes accordingly.

Mixing

Mix in a bucket or plaster tub with a D-handle drill and large paint mixer:

Mixing Sequence	Notes
¾ water	We like to put our pigment into the water, stir it vigorously, or, ideally, let it sit for a few hours to dissolve. This can prevent patches of pigment streaking in the plaster during application (though sometimes streaking creates beautiful results).
½ the sand	
wheat paste	
whiting	
clay	
pigment	If the pigment wasn't added to the initial water.
remaining sand	
water, if needed	

Coverage

1 cup (250 mL) of mix covers 2–3 square feet. If the proportions above use 1 cup (250 mL) as the unit of measurement, this will create approximately 12 cups (3 L) of mix and will cover 30–36 square feet.

Recipe

Glen's Wet-Burnish Plaster

A very smooth, very hard, but finicky earth plaster.

Contributed by Fourth Pig Green & Natural Construction

Recommended for:	Depth	Advantages	Limitations
Use over drywall or as a third coat over natural walls.	1⁄32"	Very hard plaster that will stick to anything, with a beautiful, smooth finish.	Learning curve to work with it. Applied very thin, so needs a good-quality, flat substrate.

Introduction

This plaster is not for beginners, but it can create an outstandingly smooth burnished plaster. It combines properties of an alis and fine finish plasters. The recipe was developed by Glen Byrom of Fourth Pig Green & Natural Construction after he took workshops from Carole Crews; later, Glen himself taught it at workshops in Ontario. He says he likes the 1:1:1 ratio of this recipe because it's easy to remember. It is not, however, easy to apply. At least not until you understand how it works. The high wheat paste (a third of the recipe!) makes it very sticky and hard to finish, so it needs to be very lightly misted and wet-burnished, which produces a fantastic result — if the timing and amount of water are correct. The resulting plaster is so tough that it's hard to scratch with your fingernail.

Ingredients

Ratio (by Volume)
1 clay
1 fine sand
1 wheat paste

Recommended substrates/prep

Can be used over well-prepped drywall, over well-troweled second-coat plasters on natural walls, or on other flat and smooth surfaces.

The Details

Clay — almost any bagged pottery clay will work. Kaolins are nice because of their whiteness and low silica content.

Sand — usually a fine silica sand: 80–100 grit. A fine marble/calcite sand could also work.

Wheat paste — make sure to lay plastic or wax paper onto the surface of the hot wheat paste as soon as it is done cooking. This will prevent any skin from forming that would introduce lumps into the paste. Also, be meticulous about mixing the wheat paste to prevent lumps.

Mixing

Mix with a paddle mixer or a paint stirrer on a drill. Stir the dry ingredients together, then add them to the wheat paste. Mix well, stirring for about 10 minutes.

Application and finishing details

Apply in a very thin coat (about 1⁄32 inch) using a hard square trowel. Get it as even and level as possible, but don't try to get a smooth finish on it yet — with this much wheat paste, you can't. Apply the mix to a whole wall then go back to the start with a fine misting spray bottle. Lightly mist your work in a small area (about 5–10 square feet), so that just a hint of tiny droplets shows on the surface. Pass over it with a large pool trowel; if it's still sticky, you need to mist more. If drips start running down the wall, you used a little too much water — mist less next time. You may be able to still get a good finish while following the drips down the wall, or you may have to let it dry up for 10 or 20 minutes then come back and lightly mist and trowel it again.

Coverage

The table below is for a very thin application (1/32"):

Unit of material	Coverage	Calculation
1 L clay OR 1 bag of clay	12 ft² (1.1 m²) 330 ft² (30.7 m²)	Wall area/12 = ___ L clay Wall area/330 = ___ bags clay
1 L sand OR 50 lb bag sand	12 ft² (1.1 m²) 170 ft² (15.8 m²)	Wall area/12 = ___ L sand Wall area/170 = ___ bags sand
1 L wheat paste	12 ft² (1.1 m²)	Wall area/12 = ___ L paste

Recipe

Finish Clay Plaster with Shredded Paper or Cellulose

A luscious trowel-on finish coat.

Contributed by Liz Johndrow of Earthen Endeavors Natural Building

Recommended for:	Depth	Advantages	Limitations
Finish over a clay-based level coat or as a veneer over drywall. Not suitable for uneven surfaces, as materials are quite fine, and plaster is thin.	⅛–¼" (3–6 mm)	Fast to mix and apply. A good choice for clay soil or bagged pottery clay. Paper fiber is not particularly visible in the finished plaster.	Sourcing non-ammoniated cellulose (such as Igloo cellulose) or prepping paper pulp.

This is a great recipe for preventing cracking in a high-clay plaster. The paper adds fortitude, and, depending on the amount of paper, an interesting texture. It is a replacement for other types of fiber such as fine straw or horse manure. The paper fiber allows for a slightly higher clay content for a beautiful polished finish. It makes the plaster extra creamy and increases the pleasure of troweling it on. With a lesser amount, you may not even notice it, and with an increased amount, you can create an interesting texture that has a soft feel. You can use either clay soil or bagged clay. The recipe below uses a bagged clay for simplicity of explanation. Clay-based soils always require more experimentation, as they vary so widely.

That said, a soil-based plaster is a wonderful thing to work with — if you want to experiment. It will require testing for its proportion of clay to sand and silt, as well as the strength and expansiveness of the clay. I recommend making several plaster samples and finding what works with your particular soil. Very generally speaking, ratios are 75% of a well-graded aggregate to 25% clay.

I often don't use wheat paste if I am using a clay-based soil because dusting isn't an issue if a wall is burnished, but this is an individual choice, and samples are important. A good burnishing of the plaster will prevent dusting with most soils due to the strength of the clay — unless they are particularly high in silt content (which sometimes fools people and results in a dusty finish).

Bagged clay can be a great choice if it is easier to access than clay-based soils, and it provides predictability. In this case, wheat paste or some similar starch binder will be necessary for a dust-free plaster.

If you are going for more of a veneer-type plaster with a thickness of ⅛" or less, you might even want to replace the cellulose with toilet paper. Yes, you read that right. Of course, you will want to use a post-consumer recycled paper product (not to be confused with a post-user paper product). Just soak a roll (minus the cardboard insert) and give it a quick paddle with the drill.

If you have access to lots of shredded paper, that is also an option. Soak the paper in a bucket or barrel and take the drill and a paint paddle to it the next day. It can add some interesting elements to your plaster, depending on color of paper. I would stick to non-glossy paper, though.

Application and finishing details

Apply evenly and return for a light burnish while "green" to compress and polish, either with a finish trowel or plastic.

Ingredients

Ratio (by Volume)	Quantity
0.75 kaolin or ball-type pottery clay	9.6 L
1 sand	12.8 L
0.25 whiting or marble dust — optional for smoother, finer plaster	3.2 L
0.25 wheat paste (or .05 pre-gelatinized starch)	3.2 L (640 mL starch)
0.30 ammonia-free cellulose or paper pulp — you can add more or less with sampling	4.25 L
0.5 water (may vary significantly depending on moisture in paper pulp). Start with 0.5 and slowly add more as needed. Usually up to 1x the clay content	6.4 L

Mixing

Mix in a 5–15 gallon bucket with a ½" drill and paddle. Or in a mortar tub by hand.

Mixing Sequence	Notes
½ water	
pigment, if using	Keep suspended in ½ water while adding clay. Up to 15% of binder is a general rule.
bagged or site clay	
whiting	If applicable to your recipe.
sand and dry wheat starch	Mix these two ingredients dry and mix well to avoid gelatinous clumps from the starch when it becomes wet.
cellulose	You can soak it in some of the water if using dry wheat starch or in the wheat paste if using homemade paste, or add it slowly dry so it doesn't clump.
water, if needed	You want an easy-to-spread consistency that will sit nicely on the hawk without running off. Run a margin trowel through the plaster and it should collapse a little, but not completely.

Recipe

Polishing Clay Plaster

A specialty plaster that takes a high polish.

Contributed by Tom and Satomi Lander, LanderLand

Recommended for:	Depth	Advantages	Limitations
Plastered straw bale walls, or other natural walls or plaster boards.	Paper-thin	Real natural beauty. Elegant mirror look.	Weak, compared to other plasters.

Introduction

This is a recipe for a polishing clay plaster that you can apply over tight, flat, plastered walls. This plaster is not for moisture zones like bathrooms or kitchens where water might splash. We don't recommend this mix for plastering your whole house.

Prepare a nice flat wall. Make sure it is clean — no dust or grit.

Ingredients

Ratio (by Volume)	Quantity
1 sifted clay, #200 sieve	1 cup
¼ whiting or chalk powder	¼ cup
½ water (may vary significantly depending on humidity)	½ cup

The Details

Sift your clay with #200 sieve, don't push/force the clay to get through the mesh, just shake the sieve.

Mixing

- Wear safety glasses and mask.
- Mix clay, whitening, and clean water in a clean container.
- Create a creamy mix — adjust water ratio.
- Paddle mixer and an electric drill is best.

Application and finishing details

Apply the mix paper-thin with a flexible steel trowel; spread quickly on the wall. Start polishing immediately with a clean trowel, using the whole blade to polish. When the surface starts to dry, switch the steel trowel to plastic.

Limit the size of the wall, or have several skilled plasterers working at once. Keep your trowels clean with a towel.

Practice before you apply to a real wall.

Coverage

Assuming a paper-thin an application depth, coverage is about 70–80 ft^2/1 quart.

Recipe

Starch Paste

A simple wheat paste recipe.

Contributed by Camel's Back Construction

- 10 L boiling water
- 7 cups flour
- 2 L cold water

In a separate container, mix 7 cups flour with 2 L cold water (blend together with a drill and paint stirrer, or preferably a hand blender).

Once 10 L of water is at a rolling boil, add flour/cold water mixture, mixing (with a drill paddle or hand blender) the entire time. As soon as all of the flour/water has been added and mixed into the boiling water, turn off heat and remove pot from heat to avoid burning flour paste.

Wheat paste should have a wax paper or vapor barrier seal put onto it right away to prevent a skin from forming.

Recommended to sieve it prior to using — at least the first batch, until you are certain it is free of lumps.

Recipe

Rice or Corn Starch Paste

An all-purpose recipe.

Contributed by Carole Crews

- 1 boiling water
- ¼ rice flour or cornstarch
- ¼ cold water

Blend cold water and rice/corn powder until free of lumps, stir into boiling water, continue stirring briefly until it thickens, remove from heat.

Premixed (Bagged) Plaster

This book focuses on DIY plastering, with plenty of recipes to get you started, so do you need premixed plasters? Maybe not, but they have a few advantages you should consider before you rule them out. Premixed plasters are usually excellent, and they can give you a baseline as well as something to aspire to in your own mixes. If you're only planning on adding a couple of accent walls in your house, buying premixed plaster may make more sense than making your own. See Appendix 2 for companies that offer bagged plaster.

Along with its excellent line of bagged plasters, the American Clay company has developed a full line of pigment packs that can be added to your own plasters; it also distributes tools, primers, and other materials that can be of use to a plasterer. Some of the information they provide applies equally well to homemade and purchased earth plasters.

Chapter 7

Lime Plasters

About Lime Plasters

Controlling Environmental Conditions

LIME PLASTERS ARE BEAUTIFUL, time-honored, durable plasters. It is essential that you create the correct conditions for lime, or your plaster may fail.

Lime plasters shouldn't be applied at temperatures below 40°F (5°C) or above 86°F (30°C). We are very selective about the time of year we will do lime plaster. If it is mid-summer in our warm Ontario climate, we wait for a week when the temperatures will be in the lower 70s Fahrenheit (20s Celsius) and/or there will be a rainy stint — which can mean delaying plastering to later in the season.

Carbonation, in which the plaster absorbs carbon dioxide from the air and hardens, is an important part of the curing process of a lime plaster. If conditions are such that the plaster dries too quickly, carbonation can't fully occur; there isn't the necessary amount of moisture in the plaster to allow for the absorption. Likewise, if the plaster remains very wet, carbon dioxide cannot be absorbed, and the plaster won't cure properly until it dries somewhat (this is not a common problem, but it happens). Carbonation will continue to occur whenever a lime plaster is damp (for years), but the initial carbonation is important to forming structure in the plaster, and it needs to happen slowly — over a week, or longer, so the plaster must not be allowed to dry out completely.

One-, two-, and three-coat systems

Before applying a lime plaster, the background material must be adequately prepared: clean, solid, and with adequate key. It should be dampened to reduce suction and prevent too-rapid drying of the lime plaster.

Lime plaster can sometimes be applied in a one-coat system — but only if it is going onto a relatively flat, smooth substrate, such as gypsum board that has been prepared (see Chapter 3). The sand selected for one-coat plastering may need to be fairly fine if a smooth or polished finish is desired.

For two- and three-coat systems, lime plaster wants to be applied in successive thin coats. Coarser sand is good for base coats and external plaster, whereas finer is better for finish coats. All sand for lime plaster should be sharp and well graded. For base coats and leveling coats, 1 lime:3 (or 2.5) sand is common, while finish coats may be more like 1:2, or even 1:1. Of course, depending on the local sand you are using, this can vary. It's always a good idea to do test patches until you find a ratio that works with that particular sand — in that particular climate. Hydraulic lime ratios vary depending on the type of hydraulic lime you are using (and the substrate), but generally, they are in the range of 1:2.5.

Lime plasters can be applied by hand, by hawk and trowel, by sprayer, or by pump. We prefer applying lime plasters by trowel to avoid the atomized lime particles that inhabit the air when spraying. Whatever method you use, remember to keep the coats of plaster thin: ⅜" (10 mm) maximum. If you are plastering an uneven surface, such as a straw bale wall, it is best to fill in low spots and hollows with a fiber-rich plaster before starting to plaster the wall. If you apply lime plaster too thick, it may slump; it will definitely crack, and it may not cure properly.

Hydrated lime has to absorb carbon dioxide in order to cure properly, and if it is too thick, it's possible that some of the plaster furthest from the surface may not ever adequately carbonate. For this reason, lime-stabilized earth base coats can be preferable to lime-sand as a base coat on uneven walls.

Here is a traditional three-coat lime system (other systems are described in the recipes later in this chapter):

First/scratch/pricking up coat

When applying lime plaster, work it as little as possible, as overworking it can cause slumping or cracking, or it can affect adhesion. A darby is sometimes used as a leveling tool on the first or second coat of plaster — it makes quick work of showing where the high and low spots are.

The first coat of plaster needs to be scratched or lightly textured to allow a good key for the second coat. If the scratch coat is too deep, though, it may show through on the next coat, so don't overdo it!

Fig. 7.1: *A specialized tool used more commonly in Europe than in North America, a darby makes quick work of leveling a wall.* CREDIT: MIKE HENRY

Hydrated lime (putty or powder) should cure until "green" to the touch — not soft, but you can still scratch it with a nail. It can take a week or longer before it is ready for a second coat, and some plasterers feel it is important to wait at least three weeks between coats to maximize carbonation. Pozzolans can help speed up the cure of a hydrated lime (see "Pozzolans" in Chapter 2). Natural hydraulic lime has a much faster set time — as quick as 24 hours. Regardless of the type of lime plaster, it will need to be well protected from the elements in between coats of plaster and while curing.

Straightening/leveling/floating coat

Before applying the second coat of plaster, the first coat must be misted enough that the base coat won't suck the moisture out of the new coat. This second coat will be your *leveling coat,* or *floating coat,* filling in the slight voids that are visible after the first coat. Once again, this coat should be thin (⅜" [10 mm]). Once this coat firms up somewhat (but not too much!), you can take a wooden trowel and scour (compress) the wall, working it in a circular pattern. Scouring helps seal shrinkage cracks that open up as the plaster cures, in addition to consolidating and hardening the surface. Make sure to scour all parts of the wall equally. If you scour the wall too early, you will get bubbles in the plaster, or hollows. If you wait too long, you will burnish the wall (possibly leaving black marks on the wall). Brush water lightly onto the wall as you scour. Again, this coat needs to be lightly scratched and adequately tarped from the elements. You may need to mist it while it cures.

Finish coat

The third coat of plaster is the finish coat, and it is applied quite thin. This coat of plaster can be richer in lime, as much as 1:2, or even 1:1 for very thin plaster coats on flat walls. The sand for this coat should be pretty fine. It is applied with a steel trowel, and then finished according to your taste. In traditional plastering, this finish coat would have been applied in one or two consecutive coats, each about 1⁄16" (2 mm) thick, but we tend to do our finish coat in one coat.

A rough wooden float, followed by a sponge, will give a rougher, open-grained finish — excellent for exterior finishes. A smoother wood float results in a suede-like texture with an open grain. For a smoother wall finish, a plastic float works. A steel trowel gives the smoothest finish, closing the grain on the surface. A lime plaster can be burnished (polished) by scouring or polishing the surface quite hard (while flicking lime water onto the wall). This leaves a beautiful, smooth finish. The most sensual finish of all is created with the techniques of *tadelakt*, a Moroccan plaster technique whereby after the wall has been burnished with a steel trowel, hard, flat rocks are used to really compress the surface, rubbing black soap into the wall. A recipe and instructions for tadelakt are given later in this chapter.

Curing and Aftercare

Lime plasters are pretty delicate both in application and in the conditions required for proper setting. As with all exterior plaster work, it is best to have a well-tarped building to keep the elements from drying the plaster too quickly. Some builders staple burlap to the fascia boards on the exterior, close to the plastered wall, and keep them damp to aid in slowing the cure. Our preference is to keep tarps on the exterior of a newly plastered lime wall for at least a couple of weeks, but a month or longer is better, and the walls should be regularly misted for the first week.

If you encounter problems with a lime plaster, see the Troubleshooting Guide at the end of Chapter 4.

Heritage and Specialty Plasters

In North America, heritage plasterers are getting more difficult to find. Once upon a time, plastering was a skill that was passed on from one generation to another, but with the advent of gypsum board, the old ways of cladding interior walls with wooden lath and plastering them has all but disappeared from our modern building world. Heritage plasters would more likely have been made from quicklime or hydrated lime, not hydraulic. In doing responsible restoration, it is important to match the original materials used. If you use a plaster or mortar that is too hard, you could run into failures or damage underlying materials. Cement should *never* be used for restoration work.

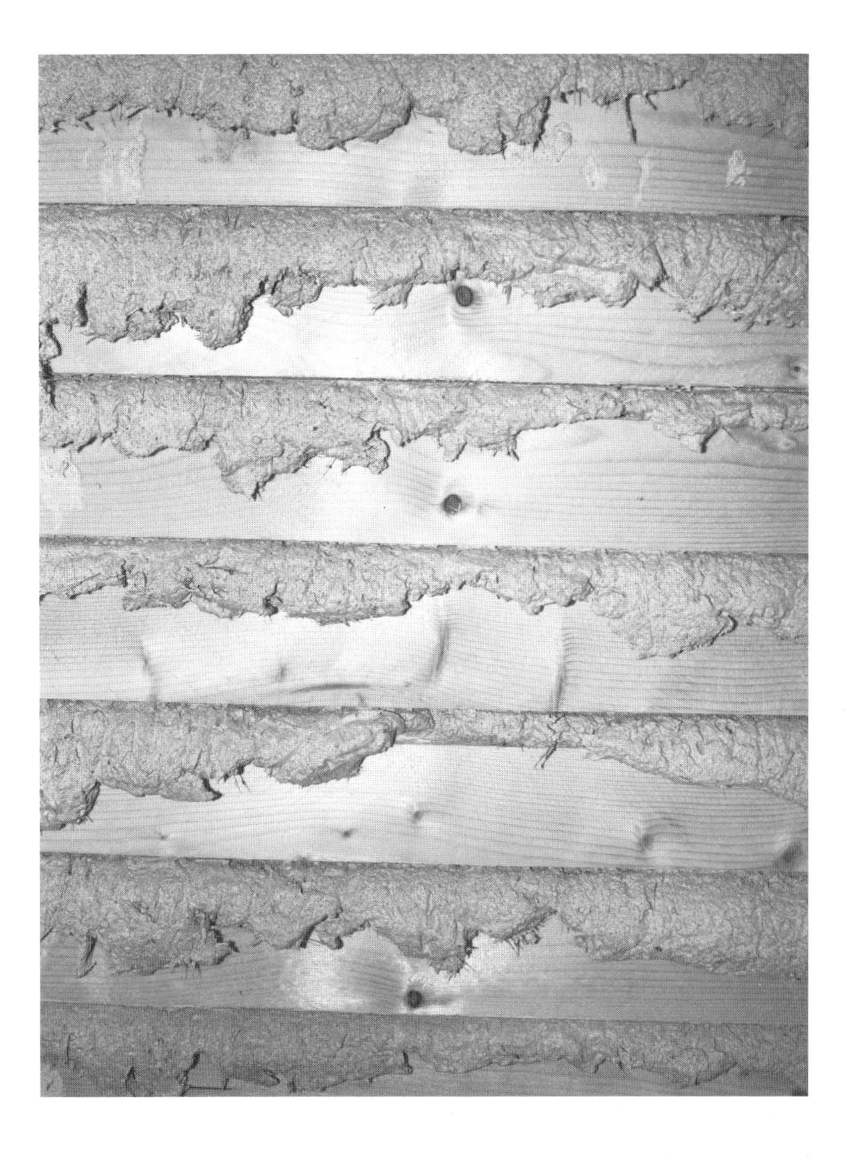

Fig. 7.2: *The nibs that key through the gaps between wood lath make the lath and plaster structurally very stable.*
CREDIT: MICHAEL HENRY

Plastering over lath would have been done with either a lime/sand/horsehair mix, or a lime putty/gypsum plaster in three coats. Lime/gypsum would be more appropriate for two-coat plasters. Gypsum isn't appropriate for exterior applications or in damp areas.

In the early 1900s, rock lath came into fashion. This was a sheet of gypsum plaster board covered with heavy paper, impregnated with holes into which the plaster would key. Gypsum board gradually replaced wooden lath, as it could be plastered with only two coats of plaster, saving time and materials.

Decorative molds, medallions, and cornices

Architectural features in North America and around the world once included such delights as plastered cornices, gargoyles, and ceilings. We have all but lost this art. Modern builders simply purchase foam or medium-density fiber board molds for ceiling rosettes and then either paint or plaster them. Master plasterers would have used a combination of lime and a gauging plaster (gypsum) to sculpt with, or to run baseboards and decorative crown moldings, cornices, and other accents. Lime and gypsum together can be cast, stamped, stencilled, or tinted. Many ceiling molds would have been done *in situ,* others would have been cast off-site and then installed.

Pargetting and ornamental decorations

English half-timber houses used to employ a technique called *pargetting* in the plaster. This method involved stamping patterns into the soft plaster to form a repeating design. Carved elements such as flowers, heads, etc. were oft times carved into plaster. Artistic flourishes on a building left a long-lasting impression. Historically, pargetting was used interchangeably with the word plastering, or *plaistering,* especially for exterior plasters.

Fig. 7.3: *Beautiful patterns can be revealed by using different colored clays and specialized carving techniques, Athena Steen's specialty.* Credit: Athena Steen

Harling, or roughcast

Harling is a type of plaster made of a very wet slaked lime to which coarse sand and gravel or stone chippings have been added. When the base coat of lime plaster is still wet on the wall, the roughcast coat is thrown against it. Certain towns had craftspeople who specialised in harling. There is a knack to learning how to flick the gravelly plaster onto a wet wall and have it look good. (See Harling recipe, below.)

Italian plasters: Venetian, marmorino, sgraffito, frescos

Some types of Italian plasters (*Venetian* and *marmorino*) involve mixtures of finely sieved marble dust (very finely screened limestone) with fine lime and sand (or lime putty) to make a plaster that, when applied in thin coats, and then burnished, or polished with a steel trowel, results in a very smooth polished wall with some mottling that looks a bit like marble.

Sgraffito involves scratching a design into plaster. The individual coats of plaster are well scoured, and often are different colors, so that when the design is scratched through the top coat, a different color will show off the intricate pattern.

We would be remiss not to mention *frescos* (the Italian word for *fresh*), the art of applying fresh pigments onto a wet plaster. The result can be beautiful paintings — such as the Sistine Chapel frescos. These pigments penetrate into the plaster, which absorbs the vibrant colors. The technique is described briefly in Chapter 9.

Lime Recipe

Simple Hydrated Lime Plaster

A lime plaster that can be mixed with materials available at most masonry supply stores.

Contributed by Camel's Back Construction

Introduction

This is not the Cadillac of lime plaster recipes, but its big advantage is its simplicity — it has just three ingredients, and all three are widely available in North America. It works fine for interiors or for exterior walls with moderate exposure, if a silicate paint is used as the final finish (one of the most durable finishes). It's not the best option for windy, exposed sites.

Recommended for:	Depth	Advantages	Limitations
Interior or sheltered exterior straw bale or other walls. May be used for finish and/or base coat up to ⅜" thick.	¼-⅜" (6–10 mm)	All materials can be purchased at most masonry supply stores. Simple to mix. Works well in many situations.	Not the strongest or most weather-resistant lime plaster that's possible.

Recommended substrates/prep

We usually use this as a finish coat, applied about ¼-inch thick over a scratched lime-stabilized earth base coat. This same mix can be used as a base coat, but a three-coat system would likely be needed due to the ⅜-inch depth limitation. We always wait a week or more between coats.

The Details

Lime — We prefer Type S bagged dry hydrated lime, which is widely available in North America and makes a good-quality plaster. Type SA is Type S lime that has additives for air entrainment — it's not what we look for, but, due to availability, we have used it in this recipe. Presumably, more entrained air would make the finish more porous.

Sand — Masonry sand (sharp, well-graded sand, up to ⅛ inch).

Fiber — polypropylene (poly) fiber, a very fine *stealth* fiber available at many masonry supply stores. Stealth fibers are so called due to their ability to almost disappear into the mix. Fiberglass should also work, but we prefer the poly.

Ingredients

Ratio (by Volume)	Quantity
1 Type S hydrated lime	50 lb (22.6 kg, 35 L)
2.5–3 masonry sand	28 gallons (105 L)
1 ounce polypropylene fiber	handful
water	

Mixing

Mix in a mortar mixer. Start with water, then half the sand, break in the bag of lime, add remaining sand. Add fiber incrementally, teasing it off from each handful and rubbing between fingers to break up clumps.

Application details

We apply and level it well using a hard rectangular trowel, then return to finish it with a pool trowel after a setting-up period (tens of minutes to hours, depending on conditions). Base coats can be applied up to ⅜ inch; finish coats more commonly at about ¼ inch. It can also be scoured with a wood float.

Finishing techniques

Finish with a pool trowel at leather-hard stage, or sponge with a slightly damp sponge, if a sandy finish is preferred. If cracks are noticed in the first few days, they can be closed with a trowel. If the wall is starting to dry, rehydrate cracks before closing them to prevent crumbling or overburnishing of the plaster. If plaster is burnished too hard, it can turn black and polished in those spots, which may not be the look you desire.

Coverage

Approximate coverage of ¼–⅜"

Unit of material	Coverage	Calculation	For 1 ft^2
1 yard sand	830 ft^2 (77 m^2)	Wall area/830 = ___ yards sand	.0012 yard
1 bag lime	118 ft^2 (11 m^2)	Wall area/118 = ___ bags lime	.008 bags
1 lb fiber (0.45 kg)	1900 ft^2 (176 m^2)	Wall area/1900 = ___ lbs fiber	.0005 lbs

Lime Recipe

Traditional Lime Putty-Based Scratch Coat with Hair Reinforcement

A workable lime putty base coat, reinforced for application over wood lath.

Contributed by Benjamin T. Scott, The Lime Plaster Company

Introduction

The use of traditional lime putty mixes can be traced back to parts of early Europe including Italy, France, and the UK. Through the years, there have been many modifications to these mixes by various cultures around the world, depending on the resources available to them and the needs of the job itself. This particular lime putty-based mix contains high-calcium lime putty, sand, and animal hair; it is one that is used extensively in plasterwork in Devon, southwest England. It should be noted that hydraulic limes may be used in place of the putty in similar ratios; however, the non-hydraulic limes we used in Devon are favored due to their superior workability and performance.

The mix itself will consist of lime and sand in ratios ranging from 1:1.5 to 1:2.5 (lime:sand), depending on the requirements of the wall or ceiling base. This plaster can be troweled on nicely, with little concern about dropping excessive material because of the sticky nature of the mix and the ease with which the material lays down under the trowel over the lath. Putty mixes allow for enjoyable work, which can't always be said for hydraulic-based finishes.

The drawback of using putty versus hydraulic limes can be the curing time — especially if you're working on a tight project schedule. In my opinion, you just can't beat high-calcium lime putty work!

Recommended substrates/prep

We use this plaster mix as a base coat over wood lath projects; however, it could be used over many other substrates, such as metal lath and masonry. The addition of water is minimal, if needed at all, to reach the desired consistency. If you do need water, using the water that covers the putty is often best because it can enable a harder set. It is known by some as *milk of lime*.

The Details

High-calcium lime putty — This is a non-hydraulic lime in a wet state, made up of approximately 95% calcium, slaked for a minimum of three to four months. The longer the material is slaked, the better. It gives great plasticity and good strength once cured, which works well for a base coat through to finish coat. Other non-hydraulic limes could be used, such as the high-magnesium limes of North America; however, we prefer the high-calcium due to our extensive use, familiarity, and understanding of this material in the UK.

Brick and concrete sand — These can be found at most building supply yards; however, it should be noted that we have sometimes found "brick sand" used interchangeably with the label "masonry sand." The sand must be washed to remove any impurities, salts, etc. Grain size and blending is also important, as outlined in the ingredients section.

Ingredients

Ratio	Quantity
5 lime putty (preferably high-calcium)	5 gallons (22 L)
8 washed concrete sand — 3.5 mm down to dust	8 gallons (35 L)
2 brick sand, washed	2 gallons (9 L)
0.5 horse hair	0.5 gallon (2 L)

Animal hair — The hair needs to be clean from oils and tannins. Generally, mane and tail hair is used for the scratch coat, but body hair can be used for other applications. The hair we tend to use is goat or horsehair; however, ox and deer have also been used in the past. The hair needs to be cut to a length between 1.5" to 2". The hair helps to prevent excessive cracking, reinforce bridging over the lath, and also hold the keys in place while the plaster is curing.

Mixing

Mix in a mortar mixer. Put in a little water and half the sand. Add the putty directly into the mixer, then add the remaining half of the sand. Let the mix turn and check the consistency. Add water, if necessary. It will improve in workability if it sits for 30 minutes or so before being applied to the wall. The mix can also sit, tightly covered, overnight and be remixed in the morning. Care must be taken not to add too much water to the mix when you remix, as this can weaken the plaster and also lead to excess cracking.

Once the material is mixed, it is then removed from the mixer, and the hair is slowly teased into it. Using a gauging trowel or similar, you can easily mix in the hair by hand.

Application Details

Hydrate the lath slightly with water before applying the plaster. This should minimize the movement of the lath whilst the plaster coat is curing. The coating should be applied so that the face of the wood lath is covered by approximately ⅜" (10 mm) of material.

The plaster is applied by trowel (usually a hard square stainless steel trowel) across the lath. Care should be taken not to press too hard in order to form the keys at the back of the lath. The application can be done in either one or two passes, depending on the preference of the applicator, to give a consistent, monolithic coating over the whole area.

After the coating has been applied and has been left to tighten up slightly, take a *scratcher* or *scarifier* and diagonally scratch key lines into the plaster surface to create a mechanical key for the next coat.

Finishing Techniques

Once the material is applied, there are two options for finishing:

1) Leave the applied coating for 24 hours, after which you can *float in* any horizontal cracking that forms from the lath, or
2) Leave the applied coat to crack and set completely. This will be covered by the next coat — the *brown coat.* (This is our preferred method.)

Lime Recipe

Multi-Functional Hemp Lime Coating

A very sculptural base coat for use on masonry walls and wooden walls.

Contributed by Gabriel Gauthier, Artcan Turnkey Hemp Construction

Introduction

The basics for this recipe come from France and were developed by Yves Kühn, the pioneer of a hemp construction process called *Canosmose*. The main application in France is for restoration and increasing the energy efficiency of stone houses. At the same time, it is a gorgeous finish, with remarkable acoustics and a nice slick, rustic look. The possibilities for application are vast, and there are also a variety of recipes.

This recipe that I am sharing here can be used to shape any form, an inch thick at a time (round window openings or plastered trim), as a replacement for gypsum wallboard for interior walls, or to stop air leaks and get more energy efficiency out of older brick, stone, or wood buildings. Its ability to regulate humidity makes it a good choice for insulating a basement.

Recommended substrates/prep

Any strong surface with a good mechanical and chemical grip is fine to receive this hemp coating. For new wall construction, we often use the wood lath technique with ¼ inch of space in between the lath, thus avoiding having to put mesh onto a plywood wall, or priming a drywall surface. Many other natural substrates that are rough enough are good, such as stone, hempcrete, brick, concrete, cordwood, cob, and clay. Sometimes it's necessary to spray a base coat, or slip coat (looks like a pancake batter, but made of sand, lime, and clay dust or cement) onto the substrate first.

Recommended for:	Depth	Advantages	Limitations
Many substrates, including wood lath, stone, hempcrete, brick, concrete, cordwood, and cob.	1" (25 mm)	Insulating. Can be built thick, in layers applied as little as 12 hours apart.	Sourcing hemp.

Ingredients

Ratio (by Volume)
2–3 hemp shiv (without dust)
1 hydrated lime (lime ratio can be higher for finish coats)
.05 red clay dust
1.25–1.33 water (up to two parts)*
* The water varies, as the lime has to be soaked; we let it soak for at least 12 hours, and up to several months.

Normally, hemp shiv comes in compressed bags of 7 ft^3 (200 liters). Some formats are 30% smaller. With one bag of hemp, we use between two to three 50 lb (22.7 kg) bags of lime, depending on the task.

For an insulated, 1-inch-thick thick coat, use 2 bags of lime; for a base coat, use 2.5 bags; and for a finish coat, use 3 bags.

Fig. 7.4: *Hemp shiv, also known as hemp hurd, is the chopped woody core of the hemp plant. It is the key ingredient of hempcrete.* CREDIT: MICHAEL HENRY

Mixing

The best equipment for mixing is our specialized flat mixer for hempcrete, but a mortar or concrete mixer also work, on a smaller scale. First, you put the shiv into the mixer and wet it until the color is yellow-gold; progressively add lime and water to soak the lime — until you get it smooth, easy to work, and a bit wetter than needed because in the next hours it's going to develop, and the lime needs water for that. Add the red clay dust (available from landscaping suppliers) to the lime. Put that mix into a bucket or other container and cover with plastic, making sure there is no exposure to air, and let it sit for several days or weeks. You get a great workable coat.

Application details

This mix keeps well — even after more than a year, if it's kept covered with water in a place where it won't freeze. Adjust with water if it's needed, and remix to get it smooth and easy to work before you use it. We do three or four layers of 1 inch each to benefit from the insulative property of the hemp/lime plaster, with 12 hours between each coat to let it dry and firm up. To help the moisture evaporate, you can open windows, warm the room to about 60°F (15°C), or use a dehumidifier.

Many application techniques are possible. Often the fastest, given the thick consistency of the plaster, is by hand, wearing gloves. I use a 4" × 14" trowel, or a triangular one with a hawk.

Finishing techniques

I use a French trowel called *langue-de-chat* (cat's tongue) to polish the surface in order to make it more durable, crack and hole-free, and to bring out the shiv and fix it at the surface, leaving a rich texture and blended color between cream and blond.

Coverage

A mix with one bag of hemp shiv (7 ft^3) generally covers 50 ft^2 of wall.

Lime Recipe

Lime Plaster with Manure

A finish or sub-finish coat.

Contributed by New Frameworks

Introduction

This coat is our go-to for a clean, somewhat rustic finish coat on natural envelope walls as well as interior walls. It mixes and trowels easily, and can be built out up to ½" (13 mm), but ¼" (6 mm) average thickness is recommended for the lime to properly cure. This lime manure coat can be applied equally easily to base coat plaster or gypsum wallboard, or it can be applied over exterior plasterboard or a lath system for exterior applications over conventional wall sheathing. There are some more technical details to pay attention to as compared to a clay-based plaster; these are discussed below.

We will, at times, omit the manure for interior gypsum wallboard applications. The manure has some odor (though this disappears once the coat dries/cures fully), and it will tint the lime slightly to a light yellow-green-white, which may be undesirable, depending on finishing options. We especially like to use cow manure in the exterior lime coat, as it lends water resistance and strength to the lime due to the reaction between the lime and the enzymes present in a ruminant's manure. Horse manure has also been very successfully used, though its contributions are primarily crack-resistance due to improved tensile strength from undigested macro-fibers. We use limestone sand for this plaster, as we have discovered (thanks to our friend and lime nerd Ryan Chivers of Artesano, Inc.) that limestone sand seems to have a chemical and molecular relationship with the active lime that yields a strong and successful plaster. The color is also then very predictable: a soft, warm white that refracts light in that way only lime can.

Recommended for:	Depth	Advantages	Limitations
Finish coat on straw bale or other natural walls, as well as interior gypsum wall board, or as part of an exterior finish system over conventional sheathing.	⅛–¼" (3–6 mm)	Strong, durable finish coat that is vapor permeable, naturally hygienic, bright in color, and flexible to resist cracking. True lime plasters reabsorb some carbon dioxide as they cure, offsetting their climate impact.	Sourcing, gathering, and working with quality manure. Lime requires temperatures above 40°F (4.5°C) to cure, and must remain moist during first week of curing.

Ingredients

Ratio (by Volume)
1 lime
3 sand
½–¾ manure
1 water (may vary significantly depending on moisture in sand and manure)

This also allows for tinting, top-coating (i.e. paint or wash), or leaving untreated, as desired.

The Details

Lime — Bagged, dry powdered Type S hydrated lime can be found at most masonry supply yards. Much of the lime in North America has magnesium carbonate in addition to calcium carbonate, so the firing process requires a process called *autoclaving,* which makes our lime unlike the traditional limes found in much of Europe. The upshot of this is that we can use our powdered lime directly, without soaking it, as was thought necessary for many years in the North American natural building field. However, this plaster is nicer after it has sat at least overnight, as the lime does "fatten up" and blend with the manure and sand into a more cohesive product after even a few hours of sitting. If you choose to let it sit, cover it well with plastic in a bin so it is airtight; exposed lime will react prematurely with the CO_2 in the air.

Sand — We use limestone sand from a local limestone quarry. They have several grades to choose from. We use what they call *feed lime,* a ⅛" minus, fairly uniform, sharp aggregate. We also use a powdered limestone called *ag lime.* We mix the two together, in a 2:1 feed-to-ag lime ratio, and use that for our sand in the mix. If we are looking for a finer finish, the ag lime provides a smoother float on the top of the coat; if we are looking for something more rugged and durable, the feed lime is more important to emphasize in the mix. ⅛" minus mason's sand may also be used.

Manure — We have used horse or cow. Both should be fresh and free of bedding or other material. We used to slurry the manure and then add that to the mix, but have since found that adding manure without additional water into the mixer with a bucket of sand and lime in the beginning of the mix, breaks it up nicely. Spending the time to source and collect fresh, quality manure is worth it for easy incorporation into the mix. Cow manure features more micro-fibers, and it creates a stickier mix as the enzymes react with the lime and clay, which helps improve the set of the plaster. Horse manure features more macro-fibers, as the horse does not digest the cellulosic fiber in its fodder.

Mixing

Mix in a mortar mixer. We use a vertical shaft, 12 ft^3 mortar mixer. If using a horizontal shaft mortar mixer, use smaller quantities because the equipment doesn't mix as well.

Application and finishing details

This plaster is very smooth and enjoyable to trowel on and to work with, as the lime imparts a plasticity that feels good on the end of the trowel, and the manure odor changes in the presence of lime into an earthy, mineral scent that is quite pleasant.

When working with lime, there is an *open working time* that is less than with a clay-based plaster. Lime will start to set up and harden anywhere from 30 minutes to 90 minutes, depending on substrate suction, ambient conditions, and troweling method. It is important to not *overwork* the lime plaster with your trowel, as you build up heat in the form of friction this way, which can dry out and damage your plaster. It is also especially important, for clay-based or other suction-heavy substrate, to pre-wet the walls several times prior to application. Ambient conditions matter significantly as well. Traditionally in the UK, lime plaster was applied in the cool, rainy spring for a reason: lime is at its strongest and best when it cures slowly, in a moist environment. So if you find yourself ready to apply lime plaster outside on a hot, sunny, windy day with the sun beating down on an adobe wall or clay-plastered straw bale wall . . . think again. Use tarps to provide shade, avoid applying lime plaster on windy days, and control substrate suction by pre-wetting.

Mixing

Mixing Sequence	Notes
lime	Wear a good respirator mask so as not to inhale particulates, and wear eye and skin protection.
sand	Mix dry together.
½ of the water	Put in enough so it mixes, but still has some roughness.
manure	Ensure it mixes in well to avoid lumps.
remaining water	Test a trowel-full from the first batch to see if water content needs to be adjusted. If storing overnight, the lime will "fatten up" and absorb water, so expect to temper in water to the mix the following day (or whenever you use it) and mix slightly wetter than desired for application.

The setting time of lime also requires thinking differently about seams or edges and when you should take breaks. Rather than taking lunch at noon when you have an open seam of plaster in the middle of a wall, time your work so the wet seam is brought to some sort of conclusion before breaking. If you don't do this, blending that seam when you begin again will be a challenge. We have had some success with leaving that edge quite thick; then cutting it back with the trowel to connect to the fresh area. But it is difficult to not have a visible seam result from this joining after one side has set up too much.

This lime plaster can be hard-troweled or burnished when it is leather hard to close up pores and to compress the coat.

When applied on a substrate with suction, this plaster will require light, twice daily misting for a minimum of 3 days up to 1 week for the lime plaster to cure. Any small cracks that open up can also be troweled shut by applying force with a small trowel 1–2 days after application. When applying on gypsum wall board, a light mist after it has set may or may not be required, depending on drying conditions. On the exterior, we recommend finishing the lime plaster with several coats of limewash or with mineral paint, so that the paint can become the sacrificial layer to the elements while working to reduce water absorption. On the interior, this plaster can be finished with a clay, lime, or mineral paint, or left natural, for a beautiful finish.

Lime Recipe

Lime Plaster with Paper Pulp

A recipe made from quicklime with added paper pulp.

Contributed by Ian Redfern, Adobe South

Recommended for:	Depth	Advantages	Limitations
Straw bale interior or exterior plaster; gypsum board; wooden lath.	¼–⅜" per coat (6–10 mm)	Inexpensive. Ideal for soft shades and washed, or textured, look. Is proven — over millennia. Fills cracks and rough surfaces.	Need to protect your skin from the caustic lime. Can't plaster in overly hot or freezing conditions. Relatively soft finish.

Introduction

These notes are not definitive and should only be used as a guide from which to develop your own methodology and practices. They have been collected from many sources, and are well proven in those environments, in addition to being owner-builder friendly. We have used and recommended these techniques for decades. Lime putty recipes can vary depending on the quality of lime putty and sands, so they must always be viewed as a starting point.

Scratch or Base Coat

Ratio (by Volume)	Quantity (*10 L buckets for measure*)
1 part stiff lime putty	2 buckets
3 parts sharp sand (not beach sand)	6 buckets
0.25 parts of damp paper fiber	½ bucket (good double handful of pressed pulp); I tend to use a little less paper in cooler, high humidity climates where there is little wind, and more in dryer climates.
Handful of short fibers per wheelbarrow load	Handful
(0.5 parts cement [max] is sometimes added to speed the set, but is best avoided.)	

Traditional Lime Putty

To make first-class lime putty, fresh burnt lime is slaked (soaked) and stored in covered containers for some considerable time (at least 3 to 9 months). Romans matured their lime for as long as five years. You can purchase quicklime pebbles to make your own putty.

Take large (50 gallon [200 L]) industrial-grade barrels and cut off the tops (the tops will become the lids). Fill them about 75% with clean water. Gradually add the quicklime pebbles, stirring to ensure that all of the particles of lime are in contact with water. We estimate a barrel will accept about 18 gallon pails of quicklime (68 L). *Note:* This slaking process is initially very violent and generates a lot of heat and steam.

Even after the slaking (transfer of CaO to CaOH) "boil" slows down, continue stirring rather vigorously to ensure the lime/water contact is maintained. Keep at least 4 inches (100 mm) of clean water on top of the lime putty.

When completely cool and clear, and water is sitting on top of the putty, close the lids and leave for at least 90 days — preferably longer. Store in a cool, still place and prevent from freezing. Cover with straw in the winter. The greater the maturity (the longer it "steeps"), the better the putty's stickiness. After this soaking time, it is ready for use as described below.

Sand

Washed river aggregates provide a range of grit sizes perfect for plaster. For really smooth surfaces we use fine-ground "dust" sands or ocean beach-ground silica sands. I prefer a slightly coarser sand as the five coats of white wash fill many of the indentations.

Paper fiber

Paper pulp fiber is created simply by agitating toilet paper rolls in the mixer until the paper has been completely teased out to a pulp. Add a bundle of toilet rolls (the least costly ones work best; remove the inner paper tube) to a couple of buckets of water in the mixer and let it run until they break down. This pulp can be left in the wet state for several days. During this time, the cellulose begins to break down, increasing the glue characteristic, which can be observed after it has been left standing for 24 hours.

De-watering is essential to minimize bringing excess free water into the lime plaster. Sufficient water removal is achieved by wrapping the half bucket of pulp in shade cloth, placing the "bag" onto a board with another overtop of it, and then standing on it to squeeze out the excess water.

Mixing

Mixing by hand is fast and produces a better plaster than any concrete mixer will. Adding water weakens the plaster and leaves the layer prone to cracking while curing. The objective is to have the lime putty absorb carbon dioxide from the air (water acts like a catalyst) as it cures. When using a bowl mixer (mechanized cement mixer), minute amounts of limewash can be added to maintain a fluid state to ensure complete mixing. Less is best.

Application and finishing details

This mix will have an active life of about four hours, depending on the weather. Thoroughly dampen the section of bales to be worked on with a fine garden hose spray or lime water mix (lime water is the excess water on top of the soaking lime) in a plastic garden sprayer before applying the scratch coat.

Rub the scratch coat into the surface thoroughly, to give a buildup of ⅜– ⅝ inch (10–16 mm). Keep the surface from drying out by setting up a "tent" to totally cover the walls being worked on. Frequently mist the walls and the tent during the curing phase. This is essential for all seasons, not only in hot or windy drying weather, as the curing phase should take a week or more — before the walls turn white. This curing continues for many years; the walls get better over time as the lime putty carbonates back to the original limestone.

This plaster can go over any number of substrates, but we most commonly use it on bale walls.

Mixing

Mixing Sequence	Notes
lime putty + damp paper fiber	Mix lime putty and paper fiber in a wheelbarrow with a hoe.
sand	Add sand gradually to the mix into the wet pool of lime putty and paper fiber. This hoeing action is more of a chopping hoe motion, and, as it plasticizes, we may use a pushing/pulling action. Add the cement (if using) and mix thoroughly.
poly fibers	Tease the poly fibers into the mix, and mix thoroughly.

Cracks

Minor shrinkage cracks in the base coat are easily "healed" by rubbing the plaster into the crack, which replasticizes the surface that binds the original layer.

Filler, or daub coat

To wad out large depressions, the hollows can be filled out with a *cob* stuffing of plaster and loose straw well mixed together. This is rubbed into the

Middle (Brown) coat

Ratio (by Volume)	Quantity (*10 L buckets for measure*)
1 part stiff lime putty	2 buckets
2–3 parts sharp sand (not beach sand)	4–6 buckets
0.25 parts of damp paper fiber	½ bucket
Handful of short poly fibers per wheelbarrow load.	Handful

Mix as above. ½ part cement is sometimes added to speed the set, but this is not recommended.

hollow and feathered out around the edges before applying the brown coat. Use several thin layers in preference to a thick one to maintain the air contact for proper curing.

Top coat

Ratio (by Volume)	Quantity (*10 L buckets for measure*)
1 part lime putty	2 buckets
.75–1 part sharp sand (fine silica sand)	1.5–2 buckets
0.25 parts damp paper fiber (not shredded paper)	½ bucket

Mix as above. Dampen and keep the wall from drying out.

Application and finishing details

Dampen the surface by misting with regular water or use lime water in a pressurized sprayer or flicked on with a hearth brush. Lay up to a thickness of ¼–⅜ inch (5–10 mm). Pure lime plasters (without cement) can be made in bulk and kept for several weeks under polythene sheeting, as long as air can't get into the mix. This plaster may be knocked back into a more plastic state by mixing it up when ready to use it.

Application and finishing details

Lay up to a thickness of ⅛–¼ inch (3–5 mm). This rather sloppy mix can be applied with a soft kitchen brush and finished with a broom for a *waterfall* effect. A coat of fresh limewash over this top coat while it is still green will assist by slowing down the cure.

For interior finishes, do a base or scratch coat, followed by a finish coat.

Lime Recipe

Tadelakt

A highly polished, water-resistant plaster originating from Morocco.

Contributed by Camel's Back Construction, based on Ryan Chivers' recipe

Introduction

Recommended for:	Depth	Advantages	Limitations
Interior walls including backsplashes, accent walls, bathroom walls, showers and tub surrounds. Only experienced plasterers who have previously completed tadelakt projects should attempt wet areas.	3⁄16" (5 mm) Applied in thin coats.	Very water resistant, beautiful.	High skill level, and learning curve. Sourcing quality limestone sand can be difficult. It is more susceptible to staining than most tile is.

Tadelakt is a water-resistant lime plaster that originates from Morocco, where it is made using incompletely burned limestone from the lime kilns on the plateau surrounding Marrakesh. Thus, even though it is said to be *pure lime,* it in fact contains sand in the form of grains of unburned limestone. Tadelakt gets its characteristic shine and water resistance from the chemical reaction between the lime in the tadelakt mix and *black soap,* a natural olive-oil-based soap that is applied while the plaster is drying and burnished into the plaster with smooth stones. The soap reacts with the lime to form *calcium stearate,* a waxy substance that is impregnated into the upper layer of lime plaster, making it hard and very water resistant, though not entirely waterproof. Tadelakt is also somewhat vapor permeable, so if some moisture finds its way in, the plaster will be able to dry. Tadelakt is a very smooth finish, with microscopic hairline cracks adding to its exotic allure.

We make our own imitation of Moroccan tadelakt using hydrated lime (Type S) and limestone sand. This saves shipping heavy materials from Morocco, reducing both cost and CO_2 emissions.

Common substrates for tadelakt include earthen walls, lime, or cement. Over stud framing, cement board may be used as a backer, with absorbent layers added for suction — these absorbent layers are an essential part of tadelakt. This is not a beginner's plaster — even skilled trowelers will want to learn on accent walls or backsplashes before attempting a shower.

This recipe is only a starting point in learning the centuries-old art of applying and finishing tadelakt. We would recommend seeking out a tadelakt workshop in North America or Europe, or, better yet, in Morocco!

Recommended substrates/prep

Tadelakt must be applied over an absorbent substrate that is at least ½-inch deep. This could be cement-lime or hydraulic lime plasters, hydrated lime plaster (applied in two coats and allowed to cure several weeks), or even solid earth walls (cob, adobe, etc.). The base should be textured — a wood float leaves a nice coarse texture, or a sponge can be used to create texture, as long as the wall isn't too hard.

Over many substrates, such as over cement board or waterproofing layers, a scratch coat of modified *thinset* (a premixed bag of cement, sand, and a water-retaining agent) is commonly applied before the absorbent base layer, to bond it to the substrate. Use a 3⁄16 or ¼-inch notched trowel (a tiling trowel), and mix thinset a little dry so the scratch doesn't slump. For small areas that won't receive any direct water, a sanded primer or other type of bonding coat could sometimes be used, but a scratch coat of thinset is always the strongest bonding coat and should be the default. Tadelakt may be applied directly to natural plasters on bale walls etc.; however, we expect it will reduce vapor permeability of that portion of the wall significantly, so limit it to small areas on exterior walls.

Ingredients

The lime/sand ratio can vary up to 60/40 either way, depending on your sand. If you have cracking in your samples, increase the sand. If you want the characteristic tadelakt microfissuring, you might want to increase the lime slightly.

Ratio (by Volume)	Quantity
1 Type S hydrated lime	20 L
1 limestone sand (calcite or marble)	20 L
0.6–0.7 water	13–14 L
Pigment (up to 10% of lime by volume)	

The Details

Lime — We use Type S bagged dry hydrated lime, which is widely available in North America and makes a good-quality plaster. Our go-to is Ivory Finish Lime from Graymont.

Sand — The sand is key in this recipe. It should be a limestone-based sand, usually calcite. The mesh sizes can be quite variable, and they will change the working properties of the mix. Generally, mesh sizes from about 20–30 down to very fine (350) is a good range. It's very important to have some fines in order to be able to achieve the polished finish for which tadelakt is famous — around 10% fines is good. If your sand lacks fine material, whiting can be added to make up for it.

Calcite sand can be bought in bags, but it can be hard to source (it is used for cultured marble sinks, specialty plasters, swimming pools, and for mixing into animal feed). You can often buy it directly from a local limestone quarry; however, most quarries aren't interested in selling a few bags of sand, and they may have a minimum order of several hundred dollars. Sometimes they'll let you take a few bags for free. Sand from quarries typically needs to be dried and sieved before using, which can be a lot of work. The sand you get from a local quarry is sometimes harder to work with, but it usually makes a beautiful tadelakt — hard work is part of doing tadelakt.

Mixing

Mix the lime and sand together with a paddle mixer (put the lime in first to make your life easier). In another mixing tub, place a measured amount of water (around ⅓ the combined volume of dry mix), saving a little water to make a pigment paste or slurry in a separate container. Add about half the dry ingredients to the water, then the pigment mix, finally the remaining dry ingredients. Mix well and add water if needed. Aim for the consistency of a thick milkshake or thin cookie dough (it should be a bit sloppy, able to hold limited shape, but will sink down with jiggling). The amount of water is about ⅓ of the volume of the dry mix, but it varies depending on your sand, the ratio of lime to sand, and pigment.

Let it sit covered for an hour or more, preferably overnight, before using.

Application and finishing

Mist surface just before application. Usually, 3–4 times for very absorptive surfaces.

Apply in 2–4 thin coats — we use at least 3, and, more commonly, 4 very thin successive coats. Each coat should be about 1⁄16 inch. The first 2 or 3 coats are rapidly and evenly sponged on with a sponge float or applied with a wood float. The final coat is applied with a rectangular trowel and smoothed with a pool trowel. Finish it as well as possible with the pool trowel, let it set up a bit, then start working it with a Japanese trowel, burnishing the surface and filling the pinholes. Try to fill all pinholes; you can work the wall when it is harder than you'd think if you use a fast burnishing motion, but be aware of consistency of finish. If you burnish too much, you will darken an area of the wall, and it may stand out.

Once it is too hard to trowel, you can begin to rub it with a stone; or, if it is already rather perfect, let it sit a little longer and come back to it when it is no longer sticky at all. At this point, give it a good stoning, so that it has a nice shine. (Tadelakt stones are polished stones of at least hardness 7 [usually quartz] with one fairly flat side.) When it doesn't feel sticky after stoning, it's ready for soap. However, you can usually get more shine by waiting up to another hour after a first stoning before applying the soap and stoning again. Apply a light coat of soap over areas that are ready, and rub the soap in with a stone.

Different areas of the wall will be ready for soap at different times. Soap them even if surrounding areas aren't ready. The edges won't show when the wall is done, but drip lines may show, so avoid letting drips run down the wall. It should be soaped before it starts to dry and lighten in color, or it will affect the waterproofing. But don't panic, the "window" for soaping is typically one to several hours — it is usually better to soap a little later rather than risk soaping too soon. *Note:* If you soap an area of the wall too soon, it will create a milky slurry and will feel soft. Stop stoning and retrowel, then wait for a while before trying again.

If you can't do a wall all in one day, you can delay the process by covering it completely with lightweight plastic overnight. Get the wall as far along as possible before you do this. Seal the plastic well, tight to the plastered wall to avoid drying at the edges.

Whenever mix is not being used, wipe the sides of the container clean, then cover the lime with water or plastic.

Aftercare

Soap the wall 1 or more times per day for a week. When soaping, always be careful to catch any drips, don't leave excess soap — always wipe excess off with a second, dry sponge, or spread it with the first sponge. Make sure to use a clean sponge to avoid scratching the fresh tadelakt.

Cracks can be closed with a stone for weeks after application; however, it is essential that the wall be hydrated well with a soap solution before stoning. Soap the wall 2–3 times over 5–10 minutes before stoning. Stoning when the wall is too dry will scratch the wall, and the edges of the crack could crumble.

Wait one month before exposing to moisture. Apply wax after one month if desired (recommended in wet areas). Howard's Citrus Shield Paste Wax (neutral) is good and relatively inexpensive. It is a beeswax/carnauba wax blend. Wax is easy to apply — rub it on with a cloth, then after about 5 minutes, buff it well with another soft cloth. If it feels sticky while buffing, wait a little longer. Wax is optional in areas that won't have regular wetting.

When cleaning is required, use a dilute solution of the black soap. This not only cleans the tadelakt, but rejuvenates it and noticeably improves beading and shedding of water.

Avoid letting water sit on tadelakt surfaces more than necessary (plan drainage of horizontal surfaces). Don't sit soap bars directly on tadelakt; it damages the surface. Oil will permanently stain tadelakt. Some bath and shower products may contain oils.

Save a jar of tadelakt mix, sealed with a little water over it, for use in future repairs. Leave it with the homeowner. It will keep approximately forever if sealed and stored properly (no freezing!). Label it with the ratio of pigment.

Caulking around edges is possible — and necessary if water sits against the bottom edge — but if drainage is good, caulking may merely create more maintenance. Caulking can strip the surface from tadelakt, so if you use it, apply it very carefully from a small nozzle so you won't need to wipe caulk off the surface.

Coverage

Coverage of mix is about 0.8 to 1 L combined dry ingredients per square foot (this will vary a little depending on your own application techniques). That is, we add the volume of lime to the volume of sand (the volume of dry mix will be somewhat less than this after combining them).

Unit of material	Coverage	Calculation	For 1 ft^2
1 bag sand (50 lb)	31 ft^2 (2.9 m^2)	Wall area/31 = ___ bags sand	.032 bags (0.45 L)
1 bag lime	78 ft^2 (7.2 m^2)	Wall area/78 = ___ bags lime	.013 bags (0.45 L)

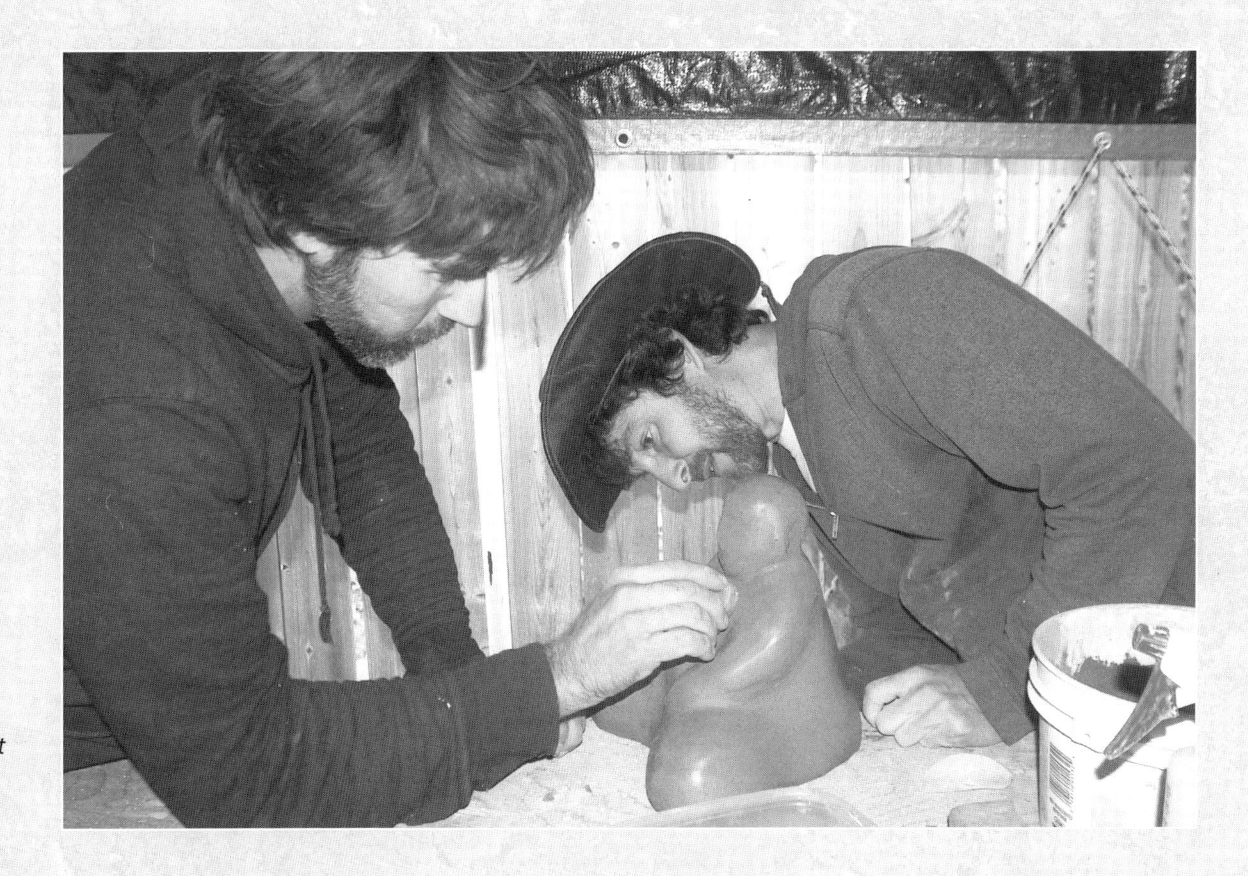

Fig. 7.5: *Tadelakt expert Ryan Chivers teaches Master Cowboy Chris Magwood the ins and outs of tadelakt on this Buddha, created during a tadelakt workshop.*
CREDIT: TINA THERRIEN

Lime Recipe

Stuc/Chevy Tadelakt

A burnished lime plaster that can have a flat polish, or a faux-tadelakt/Venetian finish.

Contributed by Camel's Back Construction

Introduction

This plaster is based on the French technique called *stuc*, but, depending on how it is applied and finished, it can take on the look of a faux Moroccan tadelakt finish. We have taken to calling this "Chevy Tadelakt." This is usually a two-coat plaster system, applied in thin (1⁄16") successive coats. Finely sieved limestone sand and Type S hydrated lime are mixed in a ratio of between 1:1 and 1:2, depending on the sand used and the desired finish.

Recommended for:	Depth	Advantages	Limitations
Interior accent walls, bathroom walls without direct water exposure.	1⁄8" (3 mm) In 2 coats.	Slightly water resistant. Attractive. Fast to apply.	Sourcing limestone sand. Skill level required varies depending on desired finish.

Recommended substrates/prep

Chevy Tadelakt can be applied over drywall that has had an adhesion coat applied to it, or over a base coat of plaster on a bale wall or other substrate. It works best on a relatively flat wall, given that it goes on in such thin coats.

The Details

Lime — We prefer Type S bagged dry hydrated lime, which is widely available in North America and makes a good-quality plaster. *Type SA* is air entrained and is not recommended for fine finish plasters.

Sand — Should be a limestone or marble sand that's not too coarse (preferably 30 mesh or finer) with good particle gradation going down to powder. If fine material is missing from your sand, add about 10–15% whiting to get a good polish.

Black soap solution — Prepare a spray bottle with dilute black soap solution (usually 20 water: 1 soap).

Ingredients

Ratio (by Volume)	Quantity
1 Type S hydrated lime	20 L
1.5 limestone sand	30 L
water	14–18 L

Mixing

Mix in a tub or large bucket. We usually use two buckets and mix dry ingredients first, then add them to the wet. We find it's easiest to just dump them in and mix all at once rather than adding gradually.

Application and finishing

For a stuc finish — an even, polished finish without much troweling character — the second coat of plaster needs to be applied soon after the first. The base coat should still be "green" (damp, and you should still be able to leave a mark with your finger-nail) when you apply the second coat. Sometimes, depending on conditions, you'll be able to apply the second coat right after the first coat goes on; other times, it will take several hours or even overnight to harden adequately. If you try to apply a second coat over a wet coat, it will slide that base coat around, and result in delamination — this isn't the right stage to apply your second coat! If you won't be able to do the second coat until the next day, take a very thin roll of plastic and gently apply it right onto the newly plastered wall (affixing it at the ceiling and floor), preventing air or drafts from entering and drying the plaster too quickly.

After we apply the second coat, we continue to trowel and smooth the walls, and apply a thin coat of olive oil soap (black soap) with a spray bottle, once it is at about the leather-hard stage. We use a steel or plastic trowel to smooth the soap into the surface. If we continue to trowel as the wall begins to harden up, a plastic trowel is important; a steel trowel will leave dark marks on any wall that is too hard.

Lime plaster and the beautiful finishes that are possible with it requires patience and a lot of keen observation to find the "perfect" time to polish a wall or to apply a second coat. Some of this will have to be learned by doing it yourself — and some of it can be shown. Interior conditions will vary in temperature and humidity, which can affect the drying and curing time of the plaster. This means there aren't set times to wait in between coats — rather, you need to babysit your walls to figure out the right timing. When you apply the second thin coat, you can make a couple of passes with a steel trowel to really smooth the plaster and bring out a bit of sheen.

To achieve a faux-tadelakt look, with more character and burnish marks, the first coat is applied a little thicker (1⁄16–1⁄8 inch) and allowed to dry completely — plan for a full day of drying. After a light misting, the second coat is applied quickly at less than 1⁄16 inch. It's important to work fast and maintain a wet edge, as it will "go off" while you're working. One or more experienced trowelers will be needed to apply and smooth it. As long as it is well leveled, some light trowel marking can be smoothed out when the black soap is sprayed on, at which point it can be burnished well with steel and/or plastic Japanese trowels. If you don't have skilled trowelers available, go for the stuc finish, which is far more forgiving.

Coverage

For coverage at 1⁄8":

Unit of material	Coverage	Calculation	For 1 ft^2
1 bag sand (50 lb)	60 ft^2 (5.6 m^2)	Wall area/60 = ___ bags sand	.017 bags (0.23 L)
1 bag lime (50 lb)	240 ft^2 (22.3 m^2)	Wall area/240 = ___ bags lime	.0042 bags (0.15 L)

Lime Recipe

Hot Mixed Lime Mortars

The revival of traditional hot-mixed lime plaster made from quicklime.

Contributed by Nigel Copsey

For plastering, a hot-mixed lime mortar delivers a material with excellent adhesion and excellent cohesion — it sticks to almost any substrate. Within the mortar, the lime, aggregate, and water are intimately combined and locked together. Used hot, the mortar will readily stiffen, but without drying out too rapidly — hot mixes are reluctant to let all their water go too quickly. Left-over mortar may be re-tempered indefinitely, so long as it is protected from drying out.

The vast majority of historic lime mortars have at least 1 part of lime to 2 parts of aggregate and often as much as 2 parts of lime to 3 parts of aggregate.

In the traditional method, when just enough water was added, a dry-mix of sand and hydrated lime would emerge. This could be stored for a period, but more commonly it was screened and mixed to a mortar for use. When the mortar was to be mixed for immediate use, a little more water might be added, but in either case, more water would be added to produce a workable mortar for immediate or prompt use.

Hot mixing methods, however, allow the water content to be readily controlled by the mortar mixer. A succession of French, British, Spanish, and North American engineers in the later 18th and early 19th centuries rigorously tested their materials and methods. All concluded that the maximum amount of sand or other aggregate that should be added to 1 part of quicklime was three parts — in the case of pure and feebly hydraulic limes. Prior to this time, lime mortars were frequently richer than 2:3. They were never leaner than this. However, if the quicklime was in the form of powder, it would leave no unslaked lumps to perform as aggregate in the mix, so 1:4 would be an appropriate maximum of sand to one part of quicklime. Pure/fat or feebly hydraulic quicklimes will typically double in volume upon slaking, so that a 1:3 quicklime:sand mix will deliver a 2:3 lime:aggregate mortar.

The temperature reached during slake should be a minimum of 100°C. If just the right amount of water is added, the temperature of the quicklime will be around 100°C or a little higher. If too little water is added, temperatures within the lime may reach 300°C. If too much water is added — or if quicklime is thrown into an excess of water — the temperature of the slake may not reach 100°C — the lime will be *drowned* and may lack binding qualities.

The volume of water necessary to complete the slake should be worked out according to the form and source of the quicklime before mixing. Just enough water will deliver a *dry slaked* mortar; a little more will deliver a thick paste. Necessary water should be delivered in one go, steadily by sprinkling or incrementally when quicklime powder is used.

Methods

There are several methods of hot mixing. A couple of these methods are described below.

Wear eye protection and dust masks and all other appropriate PPE, as for all lime (and cement) products. Have sugared water solution (Diphoterine) on hand for eyewash.

A) Mix quicklime and naturally moist aggregate at 1:3 and leave to "dry-slake" for about 3–5 minutes or until super-fine dust begins to form (rise from the mix), whether hand-mixing or mixing in a pan-mixer. Avoid using drum mixers. Use rubber containers (usually available from agricultural feed suppliers) because plastic buckets will melt. The maximum temperature at the dry-slake stage will usually be around 150°C, but it sometimes goes up to 175°C; it can also only be as low as 102°C. The temperature achieved depends on the moisture content of the sand. It should not be left to become too hot, however.

Fig. 7.6: *Quicklime pebbles are placed over damp sand.* Credit: Nigel Copsey

Incrementally add water sufficient to make a mortar of the desired consistency. Leave for 10–15 minutes before use or set aside for later use, when a little more water may need to be added during the beating. Maximum temperature after the addition of additional water and the completion of the slake will unlikely be greater than 58°C.

B) Heap moist sand and hollow the heap. Add lump or kibbled quicklime at typically 1:3 proportion by volume. Add the water necessary to effect the slake (typically about 2 volumes of water for each volume of quicklime). Modern quicklimes may be highly reactive — so much so that they will "spit" upon the addition of water. Ensure coverage of the quicklime with sand before beginning to add water. As the quicklime expands, cracks will appear in the sand covering, which will also begin to dry out. These cracks should be closed by shifting the sand with a rake or shovel to retain the necessary heat of the slake. As the quicklime slakes, continue adding water (but do not drown

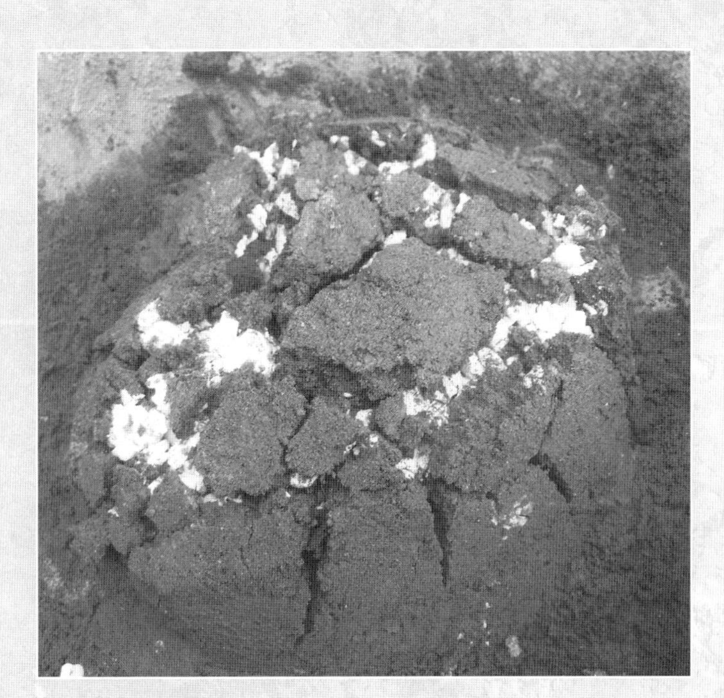

Fig. 7.7: *Cracks appear in the sand as the quicklime expands.* Credit: Nigel Copsey

Fig. 7.8: *Hotlime plaster prepared.* Credit: Nigel Copsey

or burn the quicklime) and agitating the mix with shovels. Add more water once most slaking is complete and until the mix has been brought to the required mortar consistency. Use immediately or leave for later use. The mortar should be well beaten.

Another method (using a pan mixer) won't be covered here, as most readers won't have access to a pan mixer.

Typically, for plasters and renders, hot-mix (to a mortar) the day before use, using quicklime powder, although base coats may be applied hot. The mortar will improve overnight, becoming somewhat less "tacky" and more elastic. When lump lime is used, the coarse plaster or mortar may need to be laid down for longer than 24 hours to avoid late-slaking. Some quicklimes — in either powder or lump — will require longer than 24 hours storage after mixing to avoid the risk of late-slaking.

Lime Recipe

Harling, Rough Cast, and Pebble Dash as External Lime Plastering Finishes

An aesthetically pleasing, functional finish coat.

By Andy deGruchy, LimeWorks.us

Recommended for:	Depth	Advantages	Limitations
Straw bale interior or exterior plaster.	¾" for first two coats. 3⁄16" for finish coat without pea gravel. Minimum ⅜" thick when smallest of pebbles are added.	Speeds drying time after rains due to increased surface area. Can last hundreds of years. Protects walls from elements. Aesthetically pleasing, decorative. Won't trap water in straw like Portland cement mixes can.	Certain skill to *harling*; needs to be learned. More planning required to plaster in hot or freezing conditions.

Introduction

Form follows function. Sometimes it takes many years to realize what form a construction detail should take and to incorporate the design detail into a building to let that form do its work. Unfortunately, important details can disappear in no time if time-honored techniques are not passed on to apprentices. Soon, no one realizes how profound a detail is in offering its aid to the long-term service life of surrounding building elements and the structure as a whole. This is often the case in modern times. For example, modern builders who don't know better might omit a drip kerf under a window sill, resulting in future devastating effects for a building.

This is also the case with the bumpy, textured finish that is called *harling* by the Scottish, *rough cast* by the British, and generically referred to by Americans as *pebble dash*. I have personally seen 800-year-old original harling in Scotland. However, it goes back further than that. But ask most builders about it, and you'd be met with blank stares. What makes the finish so profound as a functioning detail is that its rough and bumpy texture can actually double the surface area of a wall; this aids in drying a wet wall after each and every rain. Since we know in historic building conservation that the two main contributors to the demise of our architectural heritage are water and time (followed closely by ego and affluence), we can turn to our ancestors for instruction on how to process water better — by learning from what still stands and is working well. We can then only hope that the evidence of a technique's success falls on the ears of those who have the means to establish it once again where it should be incorporated into a design.

An obvious aesthetic reason for the form of harling, rough cast, and pebble dash is that it offers a pleasing finish that may hide some imperfections in an imperfectly flat wall. But the little-known function of these finished surfaces is that they work with nature in a symbiotic manner to process water away from the inner core of the building, which could otherwise lead to rot and decay. All properly installed lime renders can work *with* water, making it a welcomed guest rather than a relentless enemy.

Lime has interconnected pores that immediately cause moisture to spread out over the larger surface of a wall. When the surface area of a wall is increased, the speed of drying is increased. Water becomes welcomed because it happens to be a great delivery vehicle for the carbon dioxide (from the air) that the lime needs. Carbon dioxide is what properly burned limestone needs as it goes through the lime cycle that will eventually convert it back to a limestone again — after many years.

Pebble dash is especially decorative when colorful pebbles are allowed to show, having been either broadcast into the plaster or set into the render individually as *gallets*. Traditionally, pebble dash did not have a limewash coating. Pebbles in the render bulks up the substance of the render, and these tough silica bits are also what hold up strongly against weather over time — which they will do even if a limewash, once applied, is not renewed.

Recommended substrates/prep

Traditionally, harling was used over mass walls of under-fired brick and rubble stonework. The harling was a sheltering coat made from lime and aggregates. But the substrate can also be straw bale, hempcrete, adobe, slipstraw, cob, or non-rusting lath.

Ingredients

My good friend, the late Alex Hyland of the Plaster Restoration Company in Edinburgh, Scotland, liked this mix for the three coats of external lime render that finishes up with a harling topcoat.

Plaster Coats

Application details

If harling sounds like hurling, that is because you actually *do* hurl the material at the wall using a special scooped trowel or by using an old fireplace ash scoop, with a flick of the wrist. The impact of a thrown lime rendering material is profound because the bond is increased dramatically when mortar is strategically thrown with skill upon a substrate. The base coats of a lime render system can be troweled on, but if it is sprayed on or thrown on, the bond is always greater. If one were to try to push a lime render onto a straw bale substrate, the material would roll off rather than stick.

Base Coat

The base coats can be pressed into the straw or in and around lath. On flatter walls, the material is better spattered on by the means suggested. The final spattering technique of a pebble-laden render coat will broadcast the mix across the field of the wall in an even manner with an aesthetically pleasing outcome — if it is done by a person possessing good eye-to-hand coordination to guide their application. A key thing to remember is that lime renders need to cure slowly. It is always good to control suction in the substrate if the substrate is bone dry, otherwise the substrate will take the water out of the mix very quickly after it is applied and detrimentally affect its intended performance. So remember to pre-wet and then apply the harling when all dripping has ceased and the substrate is just beginning to dry out. It is okay if straw is pre-wetted because the nature of the lime render will actually draw moisture out of the straw and help the render to cure that much more slowly.

External lime render typically has three coats. Over straw bale, the total thickness can be 1½ inches or more, depending on the preparation of the bales for flatness and exactly how flat the finish has to be. The base coat is called the *pricking-up coat*.

Coat	Ratio of Lime	Ratio of Sand
Base coat	1 part NHL 5	1.5 sand
Brown coat	1 part NHL 5	2.5 parts sand
Finish harling coat	1 part NHL 5	1.5 parts sand, along with two 5 gallon (18.9 liter) buckets of pea gravel pebbles per full bag of lime

The idea is that this base coat makes the transition from whatever the substrate is to becoming part of a lime render or lime plaster finishing system. The pricking-up coat on straw bale can be ¾-inch thick; it is then scarified using a tool called a *Scotch.* This texturing step is done as the final step when installing the pricking-up coat to help increase the mechanical bond of the second coat.

The second coat on an external lime render system or an internal lime plaster system is the *leveling coat.* This coat can be ¾-inch thick or carried out in two lifts to equal 1½ inches thickness all itself. The idea of the leveling coat is that this is the coat that makes all the undulations of a wall, both up and down and side to side, become flat and on one plane. Many people, however, *like* the undulations of a straw bale wall and do not want to mask the fact that it is not perfectly flat; they would rather enjoy the waves that remain in the walls. So, the leveling coat could be just one coat used to add substance to the sheltering properties of the whole lime rendering or internal plastering system. Since it is the sand content that helps to control cracking, the leveling coat mix is usually very sandy, and therefore it is often referred to as *the brown coat.* It is referred to as *the brown coat* because sand — whether yellow or brown — is usually cheaper when it has some impurities and a bit of clay content, which make the color of this coat off-white.

The final coat is the *finish coat.* In the case of harling or pebble dash, the finish must be somewhat uniform in thickness. That thickness is determined by the size of the pebbles in the mix. Voysey Architecture in the Lake District of the UK used, in some cases, a roughcast with large-sized crushed stones to create the *Heavy Harl,* as I call it. For decoration purposes, the Heavy Harl texture would be very attractive used just around quoined corners and window and door surrounds on a straw bale house, with a lighter roughcast (pebble dash) finish used between the perimeter detailing.

Generally, the rule of thumb is to make each coat of lime plaster or render slightly weaker, as each layer is built up upon the last. The purpose of this is to allow the moisture to stay nearer to the atmosphere, and also the weaker mixes on top will posses a greater malleability in their ability to deflect and recover. This malleability is known as the "modulus of elasticity."

When attempting to achieve a long-term service life for the external envelope that shelters a straw bale house, choose pure lime/sand/pebble mixes that are right for your climate and have a known durability with the type of lime used.

Coverage

A simple rule of thumb is that a bag of NHL 5 (66 lbs) will cover 50 ft^2 of lime plaster or render at ¾-inches thick — when the ratio is 1 NHL 5 to 2 parts sharp, well-graded sand.

Lime Recipe

Homemade Hydraulic Lime Base Coat

A very workable trowel-on base coat using hydrated lime and a pozzolan.

Contributed by Endeavour Centre

Introduction

This mix mimics our typical high-straw clay mix, but uses a homemade hydraulic lime formulation that ensures rapid set time (12–36 hours) and is not susceptible to water damage or erosion the way a clay plaster might be.

The formulation for the binder was developed in conjunction with Poraver, the company that manufactures the Metapor brand of metakaolin we use. We first developed this hydraulic lime binder to glue together Poraver expanded glass beads as an insulation material, and then we began using the binder for hempcrete as well. It was a natural to try out as a plaster, since the advantages of a hydraulic lime have long been apparent. Unlike NHL-based plasters, the basic ingredients for this plaster are very inexpensive, using low-cost (and local) hydrated lime and even lower cost (and also local) metakaolin.

Metakaolin is the term for kaolin clay that has been fired at a high temperature. Naturally occurring kaolin in lime deposits is the crucial ingredient in natural hydraulic lime; in this case, we simply mix the two together from different sources. The temperature at which the kaolin is fired affects the reactivity with lime; the higher the firing temperature, the more reactive.

We've used this mix on straw bale walls and lath walls. We have left this plaster as the final finish on an exposed, hilltop site, and it has handled the weather very well.

Recommended for:	Depth	Advantages	Limitations
Well-prepped straw bale buildings, or other natural walls.	⅝–1½" (16–38 mm)	Quick setting time of NHL, but without the high cost. High-straw mix allows very thick single coat.	Obtaining suitable metakaolin.

Ingredients

Ratio	Quantity
1 hydrated lime	1 bag (22.7 kg)
1 metakaolin	1 bag (22.7 kg)
1–1.5 sand	
1.5 chopped straw (¼"–1")	
water to taste	

The Details

Hydrated lime — We use Graymont lime products, typically Ivory Finish Lime, as that's what is most commonly available locally.

Metakaolin — We've only used Metapor from Poraver. Metapor is a reactive additive based on amorphous aluminosilicate. Metakaolin is the most reactive mineral of all pozzolans and, together with calcium hydroxide, it forms the cement-like binder *calcium silicate hydrate* (CSH). Any metakaolin that meets ASTM D5370 — Standard Specification for Pozzolanic Blended Materials in Construction Applications should be suitable. However, some testing is recommended when blending new materials.

Sand — We use masonry sand or concrete sand. A sharp sand with diversity of particles up to ⅛ or 3⁄16 inch should work well.

Straw — We use a chipper/shredder with a ¾-inch screen to chop our straw. The smaller particle sizes allow for a lot of straw per volume in the mix.

Mixing

Mix in a mortar mixer.

Mixing Sequence
¾ water
½ the sand
lime
metakaolin
straw
remaining sand
water, if needed

Application and finishing details

This plaster mixes and applies like any other straw-rich formulation. The surface is prepared with a thin mix of binder and sand that is pressed into the surface as a *slip coat*. The high-straw mix is applied immediately over this base. It can be troweled on, but we often apply with gloved hands and then smooth it out with wood or magnesium floats.

The plaster will stay workable for several hours. We can typically get a one-coat system to look smooth and straight regardless of the wall texture below it. Steel trowels can be used to put a final finish on the plaster, or it can be scratched or combed to receive a final lime plaster finish.

Chapter 8

More Binders

Using Gypsum Plasters

GYPSUM, which is a relatively soft, fast-setting plaster (also known as plaster of Paris), might be the most widely used natural plaster in the construction industry, but it is surprisingly little used or understood in the natural building world. In conventional construction, gypsum is most commonly used in drywall, but both the drywall, and particularly some joint compounds, can have additives and impurities that are synthetic and may be toxic. However, in parts of North America, and in Europe, there is still a significant industry of plastering using relatively pure bagged gypsum products in two- or three-coat systems as an alternative to drywall. The results are considered to be superior to drywall aesthetically and in durability — but achieving high-quality finishes requires skilled tradespeople who understand the materials.

Gypsum is also important in ornamental heritage plastering. Much of the plaster molding work that we see on heritage buildings would be difficult to achieve without the use of gypsum.

Note that true gypsum products are always sold as dry powder, not wet mixed. Gypsum has a limited shelf life, usually three to six months. As gypsum ages, it sets more rapidly. Most gypsum plasters are only appropriate for interior use.

Common Gypsum Plastering Systems

Gypsum plastering systems include full-depth plasters as well as veneer plasters. Full-depth plasters, referred to in the industry as *conventional* plasters, are applied up to 1 inch deep in several coats; veneer plasters are commonly applied between 1⁄16 and 3⁄16 inch deep in one or two coats. Recommended substrates for gypsum plaster include metal lath, masonry or concrete, cement-board, or fiber-reinforced gypsum panels. It has been used on the interior of straw bale walls, and may have applications in many natural buildings (we'll get back to this later).

Gypsum base coat plasters may be applied ½ to 1 inch thick in one or two coats, depending on the substrate. In veneer plasters, base coats are applied approximately 1⁄16 inch thick. Base coat plasters are often site-mixed with sand for thicker coats. Follow manufacturer's instructions for ratios and mixing; generally, sand can be mixed into base coat gypsum plaster at a rate of 1–1.5 cubic feet (28–42 liters) per 50 lb bag, depending on the application. Some base coat plasters may be factory premixed with sand or perlite. Wood fiber gypsum plaster is a base coat plaster that incorporates fine wood fiber and can be used as-is or mixed with local sand (1:1 by weight).

Finish plasters can be applied to any good-quality base coat or directly to specified gypsum boards. Gypsum finishes are usually blended with lime and applied approximately 1⁄16 to 3⁄32 inch thick. *Gauging plaster* is gypsum that is designed to be mixed with hydrated lime or lime putty, while other finish plasters may be factory blended with lime, needing only the addition of water. A smooth finish is achieved by misting or brushing water on when it is beginning to set up, and then polishing the gauged lime. This removes trowel marks, closes holes, and leaves a sheen on the surface. To use gypsum plasters, follow the manufacturer's recommendations. You may also want to consult *The Gypsum Construction Handbook* (see "Publications: Print," Appendix 2).

Gypsum in Natural Building

Gypsum is a very vapor-permeable plaster (comparable to earth plasters) that has low cracking; it's been used as an interior plaster in some straw bale homes. The most common gypsum product used on straw bale homes to date seems to be USG's Structo-Lite (a premixed lightweight gypsum plaster with perlite aggregate). Other site-mixed gypsum base coats would likely have application in natural buildings; however, base-coat gypsum products are unavailable in many parts of North America. The best way to find out what's available locally is to start making phone calls to quality drywall (and masonry) supply stores.

A wide variety of finish coats may be used, including gauged lime with or without sand. Gauging plaster (gypsum that is designed to be mixed with lime) is available at many good drywall suppliers.

Another use for gypsum in interior plasters is as an accelerant — gypsum may be blended with earth or lime binders (but should not be mixed with cement) and can give an initial set that allows faster finishing or building layers sooner or thicker. Hydrated lime still needs time to carbonate before building too much depth.

Given that gypsum is a natural plaster that has low embodied energy, low toxicity, and high permeability, it's strange that natural builders don't use it more. One reason is that the best products for full three-coat plasters are only regionally available (e.g. on the East Coast of the U.S., particularly in Boston, Rhode Island, and Maryland). Another is that the skills and knowledge to use gypsum tend to be passed on through families or apprenticeships, so it's hard to get started with it. Rapid set means that work needs to happen quickly, so an experienced crew is needed, but that can be a good thing if you're running a natural plastering business. Cost is a little higher than other binders, but site mixing with aggregate can help lower the cost if the base coat plasters are available to you.

Table 8.1: Bagged Gypsum Products

Material	Description
Molding plaster	Pure gypsum, very finely ground for fast set time. Commonly used in molding. May be called plaster of Paris, though this term is also used to refer to any gypsum plaster.
Gypsum base coat plaster	Pure gypsum, commonly site-mixed with aggregate for gypsum base coats, coarser grind than molding plaster for slower set time. Similar to gauging plaster. Commercial examples include Red Top Gypsum Plaster, and Two-Way Hardwall Gypsum Plaster.
Gauging plaster	Pure gypsum, coarser grind than molding plaster for slower set time. Made for adding to lime for gauged plasters. Commercial examples include Red Top Gauging Plaster, and Gold Bond Gauging Plaster.
Lime/gypsum blends	Factory-blended gauged plasters. In some cases use of magnesium-rich lime can allow for a faster and harder set than site blending gauging plaster with Type S hydrated finish lime. Examples include USG Diamond Veneer Finish, Kal-Kote, Uni-Kal, etc.
Gypsum/sand blends	Factory blended gypsum-sand mixes can be used for high-strength base coats, particularly in veneer plaster systems. Examples include Imperial and Kal-Kote.
Gypsum/perlite blends	A blend of gypsum with perlite which is lightweight and easy to work with; it has somewhat insulative properties. Examples include Structolite and Gypsolite.
Anhydrous gypsum (dead-burnt gypsum)	Anhydrous gypsum is manufactured by continuing to bake the hydrous form to a temperature of 400°C, producing calcium sulphate, or $CaSO_4$. It has a slow set and dense crystallization resulting in a relatively hard and somewhat water-resistant plaster. An example is Keene's cement. Plâtres Vieujot in France makes high-quality blended plasters based on anhydrous gypsum, which are available in North America.

Mixing and Application Tips

Unfortunately, we haven't worked much with gypsum, so our knowledge is limited, but here are a few basic tips for working with it:

- Check the date on bags before using them; gypsum sets more quickly as it ages.
- Keep all containers and equipment clean. Don't use any cleaning water in your mixes; contamination with cured gypsum can accelerate set time.
- Maintain a consistent work flow and clean the mixer and tools between mixes.

In three-coat applications, the first (*scratch*) coat is scratched to receive the base coat. The base coat is finished by floating or troweling it without water to leave an open texture to receive the finish coat.

The finish coat is usually gauged lime, which may need to be worked three or four times as it sets — to level, fill holes, and finally burnish to a smooth finish with a misting of water. Timing is important.

Gypsum plasters may be difficult for beginners to work with due to rapid set, but this is also an advantage for accelerated construction schedules.

Even though gypsum sets quickly, thick coats will continue to dry for some time, and ventilation must be considered as for any other interior plastering.

The typical water ratio, should you need it, is about 2 gypsum:1 water, which will vary depending on type of gypsum, aggregate, etc. Whenever possible, just follow the instructions on the bag.

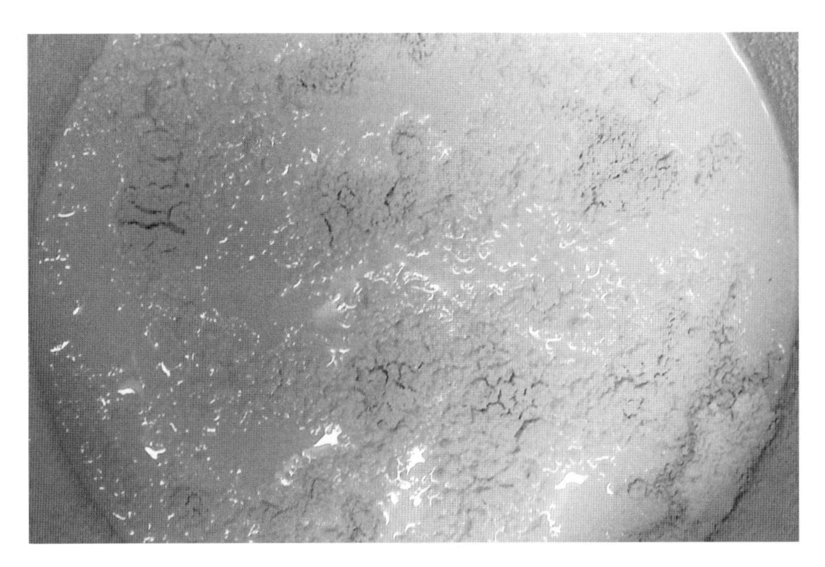

Fig. 8.1: *A good way to mix gypsum is to sprinkle the gypsum powder through your fingers into water — when it starts to form fissured islands, that's the right amount of gypsum, and it should then be stirred with a paddle mixer.*
CREDIT: MICHAEL HENRY

Using Cement Plasters

Why You Should Avoid Portland Cement

Portland cement has low permeability, which can either slow down the transfer of vapor or trap moisture in the wall. It is the most brittle of the plasters; it has little give when a building settles or moves with atmospheric conditions, making it more prone to cracking than other types of plasters. Portland cement has the highest embodied energy of all of the plaster choices — an important consideration when trying to reduce the environmental impact of your projects. Five percent of all CO_2 emissions worldwide come from the cement industry, with over 3.4 billion tons being produced annually. Some of the CO_2 (about half) comes from the high temperatures during the production of clinker; another good percentage comes from the fuel burned to attain those temperatures; and yet more is attributed to the grinding and transport of the product within the plant. In addition to the CO_2, there are heavy metals that come out of that industry, including mercury, arsenic, chromium, and hydrochloric acid, all of which damage air quality, water, plant, and animal life. Cement kiln dust is also a big pollutant. There are more stringent controls in North America now, but in some developing countries, this remains a big problem. The manufacture of lime also contributes to air pollution, but it doesn't have to burn at quite as hot a temperature, nor for as long. Lime also absorbs more CO_2 as is cures than cement does.

How to Use Portland Correctly

We don't recommend using it, but if you do choose to use Portland cement on a natural building, it must be mixed with hydrated lime to increase its permeability and flexibility.

Lime content

We typically add 1 lime to 1 cement (Type N), but more recently, when we must use cement, we have been increasing the lime content to 2:1 (Type O), to increase the permeability. Check with your engineer to see if this is okay, particularly in load-bearing situations. Holmes and Wingate note that the 1:1:6 ratio (lime:cement: sand) is better on hard backgrounds, as it is a hard mix, whereas 2:1:9 is more appropriate for weaker backgrounds. The latter mix must be very carefully batched and is only to be used in good weather conditions.

Cement/lime should be applied in either a two- or three-coat application. If plastering onto a bale wall, any large cavities should be filled (cob, hempcrete, or slipstraw are options, but they must dry fully before plastering). Gaps at bale corners may be stuffed with bundles of straw. Some experts prefer to wait a minimum of 24–48 hours in between coats, preferring to let the base coat crack in whatever way it might without transferring cracks through to the next coat. We have often applied a second coat the same day, once it is leather hard (ASTM standards allow for re-coating cement/lime plaster as soon as the base coat is solid enough to receive the next coat without being damaged). The base coat is applied to a thickness of ⅜ inch (10 mm), and then scratched. On straw bale houses, the base coat is usually used to level the wall, and the plaster may end up being quite deep (up to an inch in places). The finish coat of cement/lime is applied more uniformly, at ⅜ inch (10 mm) thick, and it can have either a fairly smooth troweled finish, or a sponged, floated finish.

The rules of good building design (adequate roof overhang, protection from splashback, etc.) are even more important with low-permeability cement plasters. Given that cement plasters have a propensity to cracking as they cure, some plasterers use expansion joints set at regular intervals to encourage the cracks to land there, similar to concrete floors.

Fibers and other additives

Additives

Additives for cement historically included ox blood, nopal cactus juice, volcanic ash, lard, and hair from hide-tanning. These additives from the past have all but fallen out of use in the modern cement world, where synthetic admixtures have replaced them. There are chemicals to retard the setting of the plaster, to accelerate it, to repel water, to increase workability, and so on. If you find yourself really needing to use a cement/ lime plaster on a building, keep in mind that the already low permeability of this plaster can be reduced even further by certain admixtures. Some of these additives can affect the strength of the plaster, or even the bond.

Fibers

Fibers must be added to cement-lime plasters to control shrinkage cracking. Polypropylene fibers, hemp, and fiberglass fibers can be successfully used with cement/lime plasters. A handful of poly fibers is generally adequate for one bag of Portland cement. About one ounce per mix is good — using more than 1–2 ounces has little if any additional benefit.

Aftercare

A cement/lime plaster must stay hydrated for several days, preferably up to a week after plastering. Gentle misting a few times a day is recommended, and it is best to keep the building tarped until the plaster has cured to reduce cracks. Cement/lime plaster will lighten in color as it cures.

Table 8.2: The Differences Between Natural Cement and Artificial Cements (Portland Cement)

	Natural Cement	Portland Cement
Chemical composition	Made from limestones that are called "clayey marls," with specific quantities of amorphous silica and aluminas.	Cement doesn't naturally contain these minerals; they are added in specific quantities during calcining.
Setting times	Sets quickly (30–60 minutes); full cure at 90 days.	Sets after a period of several hours to a day, depending on atmospheric conditions; full compressive strength at 28 days.
Flexibility, permeability	More permeable and flexible.	Not very permeable; brittle.
Mixing proportions	Rich mix: 1 cement: 1.5 sand.	Typical: 1 cement: 1 lime: 6 sand.
Application	Sets under water; typical one- or two-coat application.	Sets under water; typical two- or three-coat application.

Chapter 9

Finishes and Aftercare

Inspection

As with any type of construction, plastered walls will benefit from regular inspections. Take the time on an annual basis (and after a big storm) to inspect the plaster. Are there signs of damage? Have cracks appeared? If there is water damage, try to find the source of the problem prior to doing any repairs. Most plasters can be easily repaired once the problem is removed.

Ensure that any eavestroughs (gutters) are regularly maintained, as water flowing out of a jammed eavestrough or disconnected downspout can damage a plaster over time.

Make sure that cracks in the wall are properly repaired, as rainwater has an amazing capacity to enter a wall by capillary action via cracks. See "Repairing cracks" for earth and lime plasters, below.

Caulking

Where plaster meets with a different material, such as wood, a crack will usually form, leaving a spot for either water infiltration or air leakage. Interior cracks will result in air leaks, whereas exterior cracks can let rainwater infiltrate a wall — so exterior joints should be sealed with caulking; interior joints are sealed pre-emptively by careful use of air fins etc. Always put a bead of caulking where the base of the plaster meets flashing — this is the weakest area on a plastered wall.

Aftercare and cracks

Cracks and repairs to earth plasters

Cracks in an earth plaster can often be closed by troweling them shut while the plaster is drying, or rewetting and troweling later. Larger cracks can be opened to a "V" and filled with the plaster mix.

Repairs to unsealed plasters

Sometimes light surface marks can be removed with a pencil eraser or a slightly damp sponge. If your wall has a trowel finish, try rewetting and doing some very light surface scraping, then a light quick pass with a trowel if needed (sponging may change the reflectivity of a trowel coat, usually making the area appear lighter).

Scrapes and dings

The key to blending the repair into the existing plaster is to properly rehydrate the wall around the damage before you start. Mist several times over the course of 5–10 minutes. Use a small amount of the earth plaster mix, just enough to fill the damage. Carefully scrape the extra off.

Using a plastic trowel or yogurt lid, compress the repaired area once or twice if you need to — don't overdo this or you will burnish the wall around the repair. If it is a sponge finish, you can touch it up with a sponge.

For large repairs, it is worth letting it dry completely, then rewetting old and new mud and either retroweling or sponging.

For far more depth on this topic read James Henderson's essay in the textbox.

Repairing Cracks and Damage in Lime Plasters

Cracks that form during the initial cure (several weeks, or even longer) can often be reclosed with a trowel. Moisten the wall, let it soak in for a minute, then apply a fair amount of pressure

Clay Plaster Repairs

By James Henderson

In my time working with earth buildings and clay-based plasters, doing repair work has been the most interesting and challenging part. All walls of all houses need repairing at some point, but often it is needed just before a building is first occupied, due to the damage done to walls during the final stages of construction. The relatively soft nature of earth walls can make them more prone to cosmetic damage than lime-plastered or sheetrock walls, but they are also much easier to repair and maintain.

Working as a contractor I have repaired and restored many adobe buildings — one was over 100 years old. I have repaired earth plaster on straw bale and light straw clay houses, a few were about 20 years old. I even helped restore an 8-year-old American Clay plaster that had been installed over sheetrock in a 6,000 square foot house. In all of these repair and restoration jobs, the same techniques were used. The same techniques are also used when installing new earthen plaster. The nice thing about doing repair work is that very little new material is needed; often, the existing plaster is re-used, although sometimes the existing plaster is mixed with fresh plaster. Old adobes found on site or old piles of clay lying around the house are a huge bonus when doing this work. It is always easier to color match the plaster with the clay that was originally used. Please remember to leave bags of premixed plaster or clay with the house after doing a plaster job for the next people to be able to use.

I developed the following techniques through years of working with the gifts of clay and straw. Many amazing people and many tons of plaster have helped me come to these conclusions, which I hope you will find helpful in your work.

Clay particles are typically disc or hexagonal in shape, like little dinner plates. These disks have an electric charge that essentially make them function like little magnets. When dry, all the negatively charged sides of the disks cling to positively charged sides of another clay disk next to them. This function gives the clay a binding force. The addition of water separates the disks far enough apart to weaken the electric charge. When the water dries out, the charge is re-activated, binding the clay back together. Thus clay is in a constant state of flux; it never sets. The miracle is that when sand and straw are added to clay, it maintains a solid state that can function as a high-performance building material for centuries.

Water is the key to working with clay. How and when we use water depends on what we are trying to achieve. We can manipulate the suction of the wall and the clay content in our plaster to achieve the best outcomes. Different substrates have different suction rates; if there is no suction, the clay plaster will not sufficiently bond with the substrate. As clay does not ever set, the new plaster will physically become one with the substrate — as long as it gets sucked into it. As some of the clay in our plaster will get sucked out, we may elect to have a higher clay content in the plaster than is necessary for stabilization of the plaster to accommodate this.

Conversely, misting the substrate with water before applying the plaster will take away a lot of the suction. In this case, fewer clay disks will get integrated with the substrate, and, theoretically, you will have a weaker bond. Is this a problem when applying clay plaster to a clay wall? Hard to say, as once dry, the whole will be one. Thus, the use of water in clay plastering depends on the situation. Clay slip is sometimes used instead of mist to moisten a wall before plastering. This fulfills the same function as having a higher clay content in the plaster. Using clay slip instead of water is beneficial when the substrate is excessively silty or when applying plaster to straw bales or light straw clay. When trying to get a really strong bond for repairs or upside-down work, the substrate is not misted and the plaster is given a higher clay content than is "needed." When applying a finish coat plaster over an earth substrate, the wall may be misted to slow down ☛

the suction and give the plasterer more time to finish the plaster.

Rehydration of plaster is a technique used to repair and finish off clay plasters after they have dried. Re-hydration can be used to eliminate slump cracks and even a poor finish. Most often, rehydration is used when repairing clay plasters. The idea is to add enough water to the plaster to get it to the point where the electrical bonds loosen; then the plaster will react like it did when it was first applied. Using this approach, repairs can be made to a clay-plastered wall; then, when the repairs are dry, the area around the repair can be rehydrated or the whole wall rehydrated. Alternatively, the damaged wall area can be rehydrated first, then some new plaster is added to do the repairs. In both cases, the rehydrated area is finished upon drying in the same method as the wall was originally finished, if known. Otherwise an educated guess is made, or good old trial and error is used to try to match the original finish. In this way, the repairs will literally become part of the existing plaster and invisible. For large repairs that need a big chunk of new plaster, I typically apply the new plaster to a dry wall and rehydrate once the repair is dry. This is an attempt to get the suction of the existing wall to suck in lots of my clay binder and help dry out the repaired area more quickly. For smaller repairs and repairing silty substrates, I like to rehydrate the area first and then add the new plaster.

Rehydration is achieved through multiple mistings of the clay plaster. A water sprayer or fine hose attachment is used, and water is applied to the wall until just before it starts running down the wall. The wall only has so much suction — mist too much and you only end up with erosion. Just before erosion, move to a new area. Mist the whole area or wall and wait for 5 minutes. Then repeat the process and wait another 5 minutes. The number of times the wall needs misting depends on the clay in the plaster and the plaster mix. Three to four misting and waits are normally necessary. By the fourth time, the surface of the plaster should be soft, ideally the top ⅛ inch will move when a finger is pressed into it. At this point, a sponge or sponge float is used to reshape the plaster, before a finishing trowel is used to refinish the surface.

As clay never sets, we have the ability to use water to manipulate it. The amount of water used, coupled with the suction of the clay, allows us to maximize the benefits of using clay as a building material. No other building product is so infinitely recyclable and repairable. Once the mysteries of clay have been revealed to you, anything is possible.

while troweling. Larger cracks may need to be reclosed more than once in the first week.

Hairline cracks aren't generally of great concern. They can sometimes be filled with limewash or a finish coat of paint.

It is worth repairing cracks with a noticeable width, to avoid water wicking in from an exterior wall, and to reduce air leakage. "V" out the cracks using a grout removal tool, backerboard scoring knife, or a sharpened can-opener. Make sure that whatever you use to fill your crack is compatible with your finish coat of paint or limewash. Wet the cracks using a misting bottle, work a small amount of fine lime plaster into the crack. Use a sponge to clean up the edges. If the plastered wall is a polished, smooth plaster, use a plastic trowel to avoid black burnish marks on the wall.

Repairing Heritage Lath and Plaster

For repairs of heritage plasters, it is best to examine the plaster to determine the ingredients. Any dry plaster chunk can be separated into its base ingredients by carefully pouring muriatic acid over it, and sieving the plaster through different-sized mesh to get an accurate measure of the varied sand particles used in the original plaster. This is important when trying to conserve or replicate a heritage plaster. It may be helpful to contact an expert to help research the original plaster if you are interested in replicating it.

While hairline cracks aren't overly problematic, in the case of a larger crack, it is helpful to open up the crack to a larger area. Sometimes a premixed plaster is used to patch these cracks, one with a blend of gypsum and sand, as it only needs water. Alternatively, a high-gauge lime putty could be used (50% lime putty, 50% gypsum, or gauging plaster).

For larger repairs, you will need to use two or three coats of plaster. Cut the plaster back beyond the spot where the plaster is loose. Investigate to see if there are structural issues in the substrate — if so, repair them first. If the plaster is damaged right down to the wooden lath, remove all loose plaster and nibs; make sure that the lath is securely attached, and then hydrate this lath really well over the course of a couple of days to keep it from twisting when wet plaster (or a bonding agent) is applied. In some cases, you may wish to affix expanded metal lath to the wooden lath prior to repairing. Put on the base coat with a trowel or putty knife, and smooth it below the level of the finish coat. Once the base coat has partially set (but isn't yet dry), put on a second coat, trowel it just shy of the finish coat, and then repeat with the third coat to bring it flush to the existing plaster. When it is dry, sand it lightly and clean it with a damp sponge.

We strongly feel that when at all possible, heritage plasters are always worth saving. There are heritage plaster programs in North America and numerous opportunities in Europe for a plasterer to learn restoration skills.

Cracks in cement-lime plasters

To repair cracks, "V" out the cracks using a grout removal tool, backerboard scoring knife, or a sharpened can-opener. Wet the cracks using a misting bottle, or paint on a bonding agent. Fill cracks with grout. For most cracks, you'll want sanded grout; hairline cracks may call for unsanded grout.

Once the grout is dry, sand the area with a foam sanding block to remove excess, and make sure to wipe off any dust prior to painting with a whitewash or silicate dispersion paint. Sanding should be done on the same day, as the grout will continue to harden and become difficult to sand.

Protecting the Finish

You will want to ensure that any coating or paint that you put over a lime or clay plaster is permeable, to avoid trapping moisture in the wall.

Lime plaster will keep out a certain amount of water, but it is still a porous and permeable plaster, so it benefits from a protective coat. The lime plaster must be fully cured prior to painting it — otherwise, you may slow down the carbonation process. Hydraulic lime plaster, as previously mentioned, fares well in harsh exterior conditions, but any type of lime plaster will benefit from a protective finish coat. Any paint that is used on a lime wall must be compatible, and it must be permeable. There are lime paints that can be purchased. There is also powdered lime that, with the addition of water, becomes paint. Historically, people would wait a year for the plaster to cure before painting. In modern times, it is best to wait a minimum of three months before painting, depending on weather conditions.

Finishes

Silicate dispersion paints

Silicate dispersion paints are mineral paints, which are excellent for earth and lime walls, as they soak into the substrate and harden, forming a bond with the substrate, rather than sitting on the surface like modern paint. They are permeable, yet repel water, and are available in a wide range of natural and pastel shades. These are generally the best choice for an exterior paint on exposed walls, and they are good interior paints as well. Silicate dispersion paints were invented in the late 1800s when the King of Bavaria requested a paint that would look like Italian frescos, but endure the harsh climate of his country. These paints have good resistance to UV light. PermaTint is a North American brand of silicate dispersion paint (see Appendix 2 for suppliers).

If you are repainting a previously painted lime wall, you must ensure that the new paint is compatible with the existing coating.

Casein, or milk paints

One of the oldest paints in North America is milk-based paint. Traditionally, milk paints were used in barns, on furniture, and on homes. Milk paints would have been made from local milk and pigments (we see an abundance of iron oxide red barns in Ontario and Quebec), lime, linseed oil, and sometimes salt. Now, rather than milking a cow, we can purchase pigmented premixed powder from distributors (see Appendix 2), add water, and have instant milk paint. Of course, you can readily find recipes to make your own milk paint. As always, do lots of trials before committing to painting an entire house. Milk paints are luscious, thick paints that are easy to apply. The color palette isn't extensive, but there is a warmth to milk paint tints. There's something comforting about using a paint that our ancestors may have made for their homes. Milk paints make fine interior paints and can be used as exterior paints, but they are not as protective as silicate dispersion paints for exposed exterior walls.

Limewash

A relatively simple finish, limewash is the oldest and least expensive paint to use on a lime wall. Limewash is referred to as *self-healing* — in that it will fill in small cracks in the plastered wall, and when it rains, it may also continue to fill in cracks. A good quality limewash will protect the plastered wall, has antiseptic properties, and can be used for decorative purposes — it's an inexpensive way to brighten interior rooms. The downsides are that it generally requires more coats than most paints, and it may need to be reapplied sooner.

Limewash was historically made from quicklime, but most limewash in North America is made from a good quality lime putty (we have also made it from hydraulic lime powder). Prior to using lime putty, pour off the standing water (reserve it as part of the water for the mix). Knock up (renew) the lime putty by mixing it thoroughly with a drill and paddle. You need to dilute the lime putty quite a lot — until you get to a thin cream or whole milk consistency. (This is approximately 2 parts water to one part putty.) Put one part lime putty into a bucket or mixing container, and add the water gradually, mixing with a paddle on a drill. Depending on how wet your lime putty was to begin with, you may need up to 3 parts of water. Once you reach the desired consistency, strain the limewash through a fine sieve to remove grit or unmixed bits, and then you are ready to paint your walls.

If you make limewash from hydraulic lime, use a ratio of approximately 1 lime:3 water. Remember, you want the finished product to be like a thin cream. Mix well to make sure that the lime is well suspended in the liquid, and strain it prior to use.

Pigment can be added to the limewash; use between 5–10% of the weight of the lime, and add it only after the lime and water are well blended. Pigments must be lime compatible (see the "Pigment Characteristics" table in Chapter 2). Make sure the pigments are very fine; if they are not, run them through a dedicated spice grinder. Mix the paint periodically during the painting process to ensure that particles don't settle. Achieving a uniform color with limewash can be challenging. Careful measurements and records will be helpful in the future for color touch-ups, but when the wall is wet, there may be variations in color. It may best to mix up a large batch if using pigment in order to achieve consistent color.

Historically, other ingredients were added to a limewash such as salt, casein, linseed oil, allum, size, and tallow. Some of these additives increase the water repellency, some increase durability, while others help reduce dusting. Make sure that additives in your limewash won't affect the permeability of the finish, as you don't want to inadvertently trap moisture in your wall.

Limewash is a vibrant finish with a rich depth of color. It has been a valuable coating for protecting structures around the world against the elements.

Application of limewash

Protect windows and wood that aren't to be painted with masking. Make sure the area to be painted is brushed and free of loose particles. Dampen the plastered wall first. Masonry brushes work really well for applying the limewash, as they can hold a lot of paint. As many as 5 or 6 coats of limewash may need to be painted onto the wall for adequate coverage and protection. If there isn't pigment in the limewash, it will go onto the wall translucent, and, as it carbonates, it will dry white. Apply a thin coat of the limewash — if it is applied too thick, you will see cracking or crazing as it cures. When the first coat is dry (usually after about 24 hours) paint on another coat — continue painting successive coats until the desired color is obtained. Limewash may need to be applied again every few years. Traditionally, in barns, limewashes were routinely applied in animal stalls as a disinfectant.

Fresco

Fresco is the application of pigments to a lime-plastered wall during the initial cure, while the plaster is still damp. This allows the pigments to bond permanently to the lime. Fresco may be applied to entire walls to color the plaster, or to deepen the color of already pigmented plasters (particularly helpful for reds). It can also be used to create artwork or add artistic details to lime-plastered walls.

How to fresco

Mix lime water (the clear liquid left after lime putty has settled) with pigment. Use a ratio 1 pigment:5–10 parts lime water, depending on the pigment. Apply with a good-quality brush, stirring regularly to keep pigment in suspension.

More detailed instructions are available from the Earth Pigments Company at earthpigments.com. This site includes details on *secco* application of pigments, which occurs after initial cure, but while a lime plaster is still carbonating.

Recipe

Carole Crews' Favorite Alis

A dense, smooth alis (clay paint).

Contributed by Carole Crews

Introduction

Alis is a traditional clay paint finish that can be applied over many different substrates, including primed drywall, or as a pigmented finish over a variety of natural plasters. Alis is skin for the wall, and it can easily be refreshed with a new coating. It can be glossy or matte, smooth or textured with mica, straw, or other embellishments, and it can be any color imaginable.

Carole Crews has been instrumental in researching, experimenting, and teaching all things earth to a generation of plasterers. We are honored to be able to include her recipe for alis in our book.

Recommended substrates/prep

This recipe goes very well over a smooth absorbent surface, such as troweled earth plaster, but it may be applied over a wide variety of surfaces. Over rough substrates it could be applied in two coats; in the first coat, fine sand could replace the whiting, to help add bulk to fill the surface.

The Details

Clay — Kaolin pottery clays are recommended for whiteness and their low silica content.

Whiting — Fine calcium carbonate, available from pottery supply houses. A fine silica or marble sand may be used instead for rough surfaces.

Mica powder — Can be found online, but you'll need to hunt around a little for a bulk price. Whiting or super fine silica sand (80–100 mesh) may be used as a substitute.

Recommended for:	Depth	Advantages	Limitations
Earth and many other walls. Alternative to paint or a thin finish coat of pigmented plaster.	1⁄64" (0.4 mm)	Easy to apply, beautiful and nontoxic. Very easy to repair.	Mica can be expensive. Unsealed surface accepts stains or marks more than some paints.

Ingredients

For coverage of 150 ft²:

Ratio (by volume)	Quantity
4 water	4 quarts
5 kaolin clay	5 quarts
2 whiting	2 quarts
2 mica powder	2 quarts
1 starch paste	1 quart
Optional: slaked pigment, fine straw, mica chips	

Starch paste — May be made from cheap white flour, rice flour, or cornstarch (see starch paste recipes at the end of Chapter 6). Buttermilk is also an excellent binder; so is casein, which is easy to make from expired milk. Both are superior to starch paste, but might give off a sour milk smell while drying.

Mixing

Mix with a heavy hand whisk or with a mixing drill. Start with water, add the clay, and let it sink into the water; then add all the other ingredients except the starch paste. Mix well, then add the starch paste last, which will thicken it suddenly.

Application and finishing details

With a brush, apply onto the wall, more thickly than paint — use a low brush angle to get good coverage. A 4- to 5-inch-wide natural bristle brush is good for application, with a finer brush for edging. Apply the alis to one or two walls, then check the starting point of the first wall to see if it is beginning to dry. Smooth it with a damp sponge once it is leather hard. Rinse the sponge often and squeeze nearly all the water out before polishing the wall with a circular motion. If the sponge drags, wait until the wall dries a bit more.

Coverage

Approximate coverage for one thickly painted coat.

Unit of material	Coverage	Calculation (ft^2)	For 1 ft^2
1 qt (L) clay OR 1 bag of clay	30 ft^2 (2.8 m^2) 840 ft^2 (78 m^2)	Wall area/30 = ___ qt (L) clay Wall area/840 = ___ bags clay	.033 qt (L)
1 qt (L) whiting OR 1 lb (0.45 kg)	75 ft^2 (7 m^2) 115 ft^2 (10.7 m^2)	Wall area/75 = ___ qt (L) whiting Wall area/115 = ___ lb whiting	.013 qt (L) .009 lb (.004 kg)
1 qt (L) mica	75 ft^2 (7 m^2)	Wall area/75 = ___ qt (L) mica	.013 qt (L)
1 qt (L) starch paste	150 ft^2 (13.9 m^2)	Wall area/150 = ___ qt (L) paste	0.007 qt (L)

Appendix 1

Coverage Estimates and Conversions

CALCULATING MATERIALS FOR A JOB is not that complicated, but getting the quantities exactly right is a challenge even for seasoned plasterers. We'll start with the theoretical calculations, but this is almost always an underestimate, so make sure to see the section "Getting It Right" that comes after. Also, it's a good idea to use your test patches to confirm coverage rates.

In practice, many plasterers have a system based on multiplying the square footage of a past job by the current job, often combined with intuition. Keep careful track of your first jobs with any new plaster, because this information is invaluable in fine tuning materials ordering.

Calculating Sand

In general, the calculation starts by finding the required volume of sand. It's usually safe to assume that the binder will disappear into the voids in the plaster and will essentially occupy no space in the final plaster (unless the binder is gypsum). So for simplicity sake, let's say your plaster will be 1 inch deep; the calculation for one square foot looks like this:

> 12 inches × 12 inches × 1 inch = 144 cubic inches = 0.083 cubic feet = .0031 cubic yards
>
> *Example 1:*
> You need to plaster 1,000 square feet of wall: the calculation is 1,000 × .0031 = 3.1 yards of sand. If your plaster depth is ⅝ inch, you'll need 1,000 × .0031× ⅝ =1.94 yards of sand.
>
> *Example 2:*
> Let's say you need to plaster 30 feet of wall, 9 feet high, ¼-inch deep. Then the calculation is: 30 feet × 9 feet × ¼ inch = 5.63 cubic feet = 0.21 cubic yard.

A construction calculator is invaluable for this (available as an app for your phone) because it lets you readily multiply inches, feet, and yards or convert between them — otherwise see conversions, below.

Calculating Binder

Now, how much binder do you need? Simple, just use the ratio from the recipe. So if the ratio is 3 sand:1 binder, divide your volume of sand by 3 to get your volume of binder. Then you can use the table below to convert cubic feet into number of bags etc.

> *Example 2, continued:*
> You're making a plaster with 2.5 sand:1 hydrated lime. Continuing with the example above, if you determined you need 5.63 cubic feet of sand, since the ratio is 2.5:1 you need 5.63/2.5 = 2.25 cubic feet of binder. Since a bag of lime contains around 1.24 cubic feet of material you need 2.25/1.24 = 1.8 bags of lime.

Binder Volumes

	lb	L	Cubic feet
1 bag hydraulic lime (NHL 3.5)	50	35	1.24
1 bag Type S hydrated lime	50	35	1.24
1 bag ball clay	50	23	0.81
1 bag Hawthorn Bond clay	50	20	0.71
1 bag Redart clay	50	23	0.81
1 bag kaolin clay	50	28	1
5 gallons clay slip		19	0.67

Other notes on binder volumes:

A standard (short bed) pickup truck loaded to an average depth of 12 inches contains 1 yard of material. A full-sized pickup loaded to 12 inches holds 1.5 yards.

1 bag of lime = 35 L dry = 24 L thick lime putty when mixed with 14 L water

Calculating Fiber

In theory, the fiber will occupy space in a plaster — however unlike sand, fiber tends to be very light, with more air space than solids — so we find that for most plasters, fiber can be treated like binder, and simply calculated as a ratio of the sand that doesn't occupy significant volume. For plasters where there is more straw than sand, this may need to be rethought — even then, the fiber will occupy a fraction of the space once wet, so it can't be considered to occupy space in the same way sand does. Here are a few useful numbers for calculating fiber additions:

1 bale intact* = 120 liters or 4.5 cubic feet or 0.12 cubic meters or 0.15 cubic yards (straw bale density = 7 lb/ft^3)

1 bale chopped (coarse) = 240 L (coarse chopped straw density = 3.5 lb/ft^3)

1 contractor's bag holds around 120 L

1 L hemp sliver = 75 g

1 kg hemp sliver = 13 L

*In Canada, we always refer to 2-string bales. We have never seen a 3-string bale here.

Getting It Right: How Much to Adjust the Numbers

If you were to simply use the numbers you just worked out, you'd likely run out of materials before the job is done. At a minimum, add 10–15% for waste, filling unevenness in the wall, etc. For base coats, remember that in addition to your depth of application, quite a bit of material may be pushed into the substrate — consider adding 20% or more.

If you're ordering bulk masonry sand by the yard, you should add 30–35%, partly because of increased waste (make sure to lay out a tarp under your sand pile), but mostly because of a phenomenon called *bulkage.*The volume of sand when it is saturated (as in plaster) and when it is completely dry are very similar; however, when sand is merely *damp* it swells significantly, commonly 25–35%. This means that when your damp sand is mixed wet in plaster, it will shrink and occupy *less* volume, in the mixer and on the wall — so more sand is needed to meet the required depth.

You would usually order 5–10% less binder than the ratio of bulk sand would suggest, because there's usually less waste of binder (and it's far more costly to have extra binder than extra sand).

Example 2, the conclusion:

So far we've determined we need 5.63 cubic feet of sand, and 2.25 cubic feet (1.8 bags) of lime to cover our 30 foot × 9 foot wall with ¼ inch of plaster. This is a finish coat, and the base is well leveled; however, the sand is masonry sand that was stored outside in the yard, and it is damp, so we'll order 30–35% extra sand to account for bulkage.

5.63 × 1.35 = 7.6 cubic feet of sand, 0.28 yards, and

2.25 × 1.25 = 2.8 cubic feet of lime, or 2.26 bags of lime

Note that the lime is adjusted about 10% less than bulk sand (assuming it's reasonably easy to pick up a few extra bags if needed). If your calculations were based on dry sand, you could use the same ratio (i.e. × 1.15 for each). Take all of this with a grain of salt. As you complete jobs, you should adjust these numbers to match your experience.

A few more useful numbers

Clay slip coverage, bagged clay

1 bag of kaolin clay, mixed 1.25:1 with water and sprayed, should cover about 290 ft^2. On poor substrates, it could be mixed thicker and sprayed more heavily, and coverage might be as low as 200–250 ft^2/bag.

Starch

250 mL (1 cup) pre-gelatinized wheat starch substitutes for 1 L wheat paste
250 mL dry starch = 100 g

Conversions

1 cubic yard	= 808 quarts
	= 765 Li
	= 0.765 cubic meters
	= 27 cubic feet
	= 46656 cubic inches
1 cubic foot	= 29.9 quarts
	=28.3 L
1ft × 1ft × 1inch	= 144 cubic inch
	= 0.0833 cubic feet
	= 0.00308 yards
	= 2.49 quarts
	= 2.36 L
1 kg	= 2.2 lb

mm	Fractional Inches (Approximate)	Decimal Inches
1 mm	>1/32 inch	0.039
2 mm	>1/16 inch	0.078
3 mm	>3/32 inch	0.118
4 mm	>1/8 inch	0.157
5 mm	>3/16 inch	0.196
6 mm	<1/4 inch	0.236
7 mm	>1/4 inch	0.275
8 mm	5/16 inch	0.314
9 mm	<3/8 inch	0.354
10 mm	>3/8 inch	0.393
11 mm	7/16 inch	0.433
12 mm	<1/2 inch	0.472
13 mm	>1/2 inch	0.511
14 mm	9/16 inch	0.551
15 mm	<5/8 inch	0.590
16 mm	5/8 inch	0.629
17 mm	<11/16 inch	0.669
18 mm	<3/4 inch	0.708
19 mm	3/4 inch	0.748
20 mm	<13/16 inch	0.787
21 mm	>13/16 inch	0.826
22 mm	<7/8 inch	0.866
23 mm	>7/8 inch	0.905
24 mm	15/16 inch	0.944
25 mm	1 inch	0.983

Appendix 2

Resources

References

Publications: Print

PLASTERING REFERENCES are too numerous to print here. We used a vast bibliography of materials in researching this book, some of which include out-of-print books on lime. Watch for an electronic bibliography of all sources in the future at essentialnaturalplasters.com. Here's a list of references we cite directly and/or consider to be the most important.

Crews, Carole. *Clay Culture: Plasters, Paints and Preservation.* Gourmet Adobe Press, 2010.

Crimmel, Sukita Reay and James Thomson. *Earthen Floors: A Modern Approach to an Ancient Practice.* New Society Publishers, 2014.

Guelberth, Cedar Rose and Dan Chiras. *The Natural Plaster Book.* New Society Publishers, 2003.

Henderson, James. *Earth Render: The Art of Clay Plaster, Render and Paints.* Python Press, 2013.

Holmes, Stafford and Michael Wingate. *Building with Lime.* Rugby, Warwickshire: Intermediate Technology Publications Ltd., 2006.

Lazell, Ellis Warren. *Hydrated Lime: History, Manufacture and Uses in Plaster, Mortar, Concrete: A Manual for the Architect, Engineer, Contractor and Builder.* Pittsburgh, PA: Jackson-Remlinger Printing Co., 1915.

Minke, Gernot. *Building with Earth: Design and Technology of a Sustainable Architecture.* Germany: Birkhäuser, 2009.

Racusin, Jacob Deva and Ace McArleton. *The Natural Building Companion.* Chelsea Green Publishing, 2012.

Reynolds, Emily. *Japan's Clay Walls.* Createspace Independent Publishing Platform, 2009.

Sickler, Dean. *The Keys to Color: A Decorator's Handbook for Coloring Paints, Plasters and Glazes.* Createspace Independent Publishing Platform, 2010.

Thomson, Margaret L. "Why is Type S hydrated lime special?" International Building Lime Symposium 2005. Orlando, Florida, March 9–11, 2005. buildinglime.org/Thomson_TypeS.pdf

USG. *The Gypsum Construction Handbook: Edition 7.* John Wiley & Sons, 2014. Centennial edition is available free at usg.com/content/usgcom/en_CA_east/resource-center/gypsum-construction-handbook.html

Van Den Branden, Felicien and Thomas L. Hartsell. *Plastering Skills.* American Technical Publishers, 1985.

Weismann, Adam and Katy Bryce. *Using Natural Finishes: Lime- and Earth-Based Plasters, Renders and Paints.* Green Books Ltd., 2008.

Online Forums, Discussions, Websites

The Building Limes Forum buildinglimesforum.org.uk

Building Conservation buildingconservation.com

The Last Straw Journal. thelaststraw.org

I Love Natural Plaster, Facebook group. facebook.com/groups

Master of Plaster, discussion of wood lath plastering. masterofplaster.forumchitchat.com/post/new-wood-lathe-plaster-construction-2516144

Straw Bale Social Club conversation: Lime over clay. groups.yahoo.com/neo/groups/SB-r-us/conversations/topics/14236

Organizations

Builders Without Borders builderswithoutborders.org

California Straw Building Association strawbuilding.org

Development Center for Appropriate Technology dcat.net

Natural Builders Northeast nbne.org

Ontario Natural Building Coalition naturalbuildingcoalition.ca

Suppliers

This is only a partial list of some of the suppliers we use.

Lime

Graymont. Quicklime, lime putty, and hydrated lime. graymont.com

Lancaster Limeworks. Lime putty, premixed plasters etc., in Lancaster, PA. lancasterlimeworks.com

Limes.us. Hydraulic limes, black soap, lime paint. limes.us/tag/canada

Limestrong plaster. Pozzolanic lime plasters made from lime and pumice. A full line of plasters from coarse basecoats to Venetian style fine finishes. limestrongfinish.com

Limeworks. Premix NHL lime, lime putty, lime paint, black soap. limeworks.us

Preservation Works. NHL. preservationworks.us

Skycon Building Products, Toronto. NHL, lime, gypsum, marble sand, trowels. Toronto. skycon.ca

US Heritage Group. Lime putty and NHL lime. usheritage.com

Virginia Lime Works. Lime putty, natural hydraulic lime, natural cement. virginialimeworks.com

Clays

American Clay. Bagged clay plasters and paints from New Mexico. americanclay.com

Capital Pottery Supplies & Materials Ltd, Ottawa, ON. Carries or can order bagged clay. capitalpotterysupplies.com

Hutcheson Sand & Mixes. Infield clay mixes. Huntsville, ON. hutchesonsand.com

Mar-Co Clay. Infield clay mixes, available in the northeastern, southern, and midwestern United States and across Canada. marcoclay.com

Plainsman Clays. Raw Materials. Medicine Hat, Alberta. plainsmanclays.com

The Pottery Supply House. Raw materials, bagged clay. psh.ca

Tucker's Pottery Supplies Inc., Richmond Hill, ON. Raw materials, bagged clay. tuckerspotteryeshop.com

Natural cements

Edison Coatings Inc., Plainville, CT. Lime, natural cement, roman cement. naturalmortars.com

Rosendale Natural Cement Products. rosendalecement.net

Hemp

Plains Hemp, Alberta. Hemp hurd, hemp fibers. plainshemp.com

Ontario Hemp Materials. Hemp fibers. ontariohemp.ca

Stemergy, Delaware, ON. Hemp hurd and shiv. hempline.com

Additives, Pozzolans

Metapor Poraver, Innisfil, Ontario. Metakaolin. poraver.com/us

Pigments and soap

Ardec Finishing Products, Saint-Sauveur, QC. Pigments, black soap, casein, whiting powder, wax. ardec.ca/en

Earth Pigments, Cortaro, AZ. Pigments, Marseilles soap, whiting, mica, marble dust. earthpigments.com

Kama Pigments, Montreal. An artist-owned business, pigments, wax, oils, black soap. kamapigment.com

Kremer, Pigments Inc., New York. Pigments, wheat and rice starch powders, olive oil soap. shop.kremerpigments.com/en

Natural Earth Paint. Bulk pigments and paints. naturalearthpaint.com

Sinopia, San Francisco, U.S. Pigments and milk paints. sinopia.com

Limestone/Calcite/Marble sand: Local quarry, manufacturers, or masonry supply stores

Coloured Aggregates. Produces limestone sands. Aurora, ON. Minimum order 1+ palettes. colouredaggregates.com

Huber Engineered Materials. Produces swimming pool aggregates (Durawhite). hubermaterials.com

Imersys Carbonates — produces swimming pool aggregates. Also 40–200 mesh calcium carbonate sand, a good sand for some finish plasters (but lacks fine material for polishing). imerys-carbonates.com

Merkley Supply. Ottawa retailer, carries colored aggregates SW350 calcite sand in 50 lb bags. Also carries Natural Hydraulic Lime. merkleysupply.com

Minerals Technologies. Produces swimming pool aggregates (Marblemix). mineralstech.com

Shelburne Limestone Corporation, 688 Quarry Rd, Shelburne, VT 05482, (802) 985-2334. Limestone sand. Minimum sale may be a pickup load.

Skycon Building Products. NHL, lime, gypsum, limestone sand, trowels. Toronto. skycon.ca

Upper Canada Stone Company, Madoc, ON. Limestone sand, $300 minimum order. uppercanadastone.com

Pottery supply retailers, marble finishing businesses, cultured marble manufacturers, and pool supply retailers are also places to try to find limestone sands. Delaware tadelakt artisan Petros Dandolos recommends Marblemix (Minerals Technologies/Specialty Minerals), which he was able to buy at a reasonable mark-up from a swimming pool installation company. If you can forge a relationship with a local installer this will spare you the usual minimum order of one palette.

Finishes, paints

Bioshield Healthy Living Paints. Clay paint, pigments. bioshieldpaint.com

Homestead House Paint Co., Toronto. Milk paints, natural oils, stains, beeswax. homesteadhouse.ca

Kremer Pigments Inc., New York. Pigments, wheat and rice starch powders, olive oil soap. shop.kremerpigments.com/en

PermaTint Ltd., Concord, ON. Silicate dispersion paints. permatint.com

Tockay Natural Paints, Montreal, QC. Natural paints and plasters, pigment, wax, olive oil soap. tockay.com/en

Japanese trowels

Hida Tool & Hardware, Berkley, CA. hidatool.com/woodworking/trowels-for-plaster

Japanese Plastering Trowels, Japan. japaneseplastering.com

Landerland, Hillsboro, NM. Japanese steel and plastic trowels. japanesetrowels.com

Rising Sun Gardens, Minden, ON. risingsun-gardens.wixsite.com/risingsun-gardens

Eco-stores

Eco Building Resource, Aurora, Ontario. Nontoxic house building materials. eco-building.ca/about.htm

Living Room — Kingston: Living Rooms Ecological Living and Building, Kingston, ON. Paints and finishes, American Clay plaster. livingrooms.ws

Nonmetallic lath

BASF. PermaLath 1000. Fiberglass lath. basf.com

Plastic components. Ultra-lath plus Plastic lath. plasticcomponents.com

Spider Lath. Fiberglass lath. store.spiderlath.com

Homasote fiber board

homasote.com to find a local distributor.

Screens, filters

McMaster-Carr. Many mesh sizes and types of screens. mcmaster.com

Contributors

GLEN BYROM is a founding member of Fourth Pig Green & Natural Construction. He has over 20 years of experience in natural construction, renewable energy, and related programs. He has worked on over 30 straw bale houses, commercial buildings, and as many photovoltaic (PV) installations in Canada and the U.S. A certified Passive House builder, he has taught natural building workshops with the Fourth Pig and at Solar Energy International in Colorado, specializing in hands-on instruction in natural construction and finishing including straw bale, cob, rammed earth, earth bag, passive solar design, and adobe. Glen brings his expertise in natural and solar building techniques, renewable energy system design and installation, and a passion and creativity for beautiful, healthy, and functional spaces to the Fourth Pig.
fourthpig.org

Camel's Back Construction
The original Camels, Chris Magwood, Peter Mack, and Tina Therrien, started off as cowboys and cowgirls in the wild North, learning the ropes of plastering, inventing new recipes, and incorporating new techniques. They were fortunate enough to collect a posse of dedicated and talented plasterers over the years. Some notable top-notch Camels include the following: Mike, Karla, Heidi, Chris, Pete, Kat, Tola, Andrew, Andie, Justin, Paul, JP, Steve, Leslie, Monica, Luke, Tim, Jim, JD, Joe, Jan, Dutch, Jake, Drake, Dale, Greg, Karen, John, Annie, Julie, Ryan, Burt Sturton, Fraser, Paul, Ben, Ian, Ainslie, Dave, Denis and oh! so many more. This book would not have come to fruition without the ideas and assistance of the many, many Camels. We are eternally thankful to each and every one of you. A special shout out to Karla Holland who was mixer-in-chief for some formative years as we transitioned to earth and lime plasters, and who was involved in the development of some of the recipes presented in this book. Many thanks to Andrew McKay for sharing his earth plaster knowledge with us during the transition years. The biggest nod of all goes to the Wild & Fearless Dromedary himself, Michael Henry, whose diligence, curiosity, research, dabbling, and supreme troweling skills led the Camels through many, many happy plastering adventures. And of course, to all of our generous clients who allowed us to play with mud, thank you!

Nigel Copsey began working with stone as a drystone waller in Cornwall. After eight years, he trained as a stonemason and carver at Weymouth College in Dorset, becoming established as a lettercutter and stonemason, working mainly in the field of stone and building conservation. Based originally in the southwest, the company has carried out projects across England and more recently in Vermont and Andalusia. Nigel Copsey is a fully accredited member of the United Kingdom Institute of Conservation.
nigelcopsey.com

Carole Crews grew up in adobe houses in a place where "natural building" came naturally, having been practiced by Native Americans and Spanish settlers for centuries. She has adapted the most efficient and effective techniques and made them available to owner-builders who wish to simplify their methods and surround themselves with natural beauty. For additional information, see her book, *Clay Culture: Plasters, Paints and Preservation.* Carole also uses these materials for art projects and believes it is the creative process that makes us most human. As the years accumulate, she has turned to creating music.

Andy deGruchy is the owner of LimeWorks.us, the leading manufacturer and distributor of "green" historic preservation and sustainable building mortars, plasters, and paints in the United States. His lime-based materials are used for durable, appropriate, time-tested historic masonry restoration campaigns. In 1979, Andy received a full scholarship for a three-year program to study and practice masonry from the nation's oldest private trade school, The Williamson Free School of Mechanical Trades in Pennsylvania, which was founded in 1888; while there, Andy received their "Key" award for the advancement of its founding ideals. He maintains a few specialty crews which have operated since 1986 as deGruchy Masonry Restoration. They have restored hundreds of historic buildings in the Delaware and Lehigh Valleys of Pennsylvania. Andy gives lectures and conducts training workshops on the subject of historic masonry restoration and sustainable building.

The Endeavour Centre is a not-for-profit sustainable building school based in Peterborough, Ontario. Through full-time programs and short-term workshops, they help educate owner-builders, design and construction professionals, and building officials on a wide range of sustainable building methods and materials.

endeavourcentre.org

Gabriel Gauthier worked in France from 1998 to 2000 to learn the techniques of hemp construction and lime finishes. He created ArtCan in 2003 to design and build hemp buildings that are both more durable and more comfortable, and he has worked on more than 100 buildings since ArtCan's creation. He is a professional craftsman member of the Conseil des Métiers d'Art du Québec. ArtCan is pursuing applied research to develop new tools and products to promote the evolution of organic hemp and ecological hemp construction, and Gabriel teaches workshops on these ancient building techniques. See hempconstruction.com for more information.

James Henderson is a builder who is happiest when building with clay. He believes that clay is the highest-performance building material available. His passion for it has led to a career building straw and earth homes for clients. The same passion comes through in his book *Earth Render*. James lives in Washington State with his wife and two kids on a permaculture property. He can be contacted at naturalbuildingsupplies@ zoho.com or at Henderson Clayworks on Facebook.

Kaki Hunter and Doni Kiffmeyer have been messing about with dirt since they took their first Earthbag workshop with Nader Khalili at the Cal Earth Institute in 1992.

They set out to share their passion for building with dirt by giving workshops and authoring *Earthbag Building: The Tools, Tricks and Techniques*. They encourage all types of natural building with an emphasis on meeting the FQSS stamp of approval: Fun Quick Simple and Solid!

Please visit okokokproductions.com for more information.

Liz Johndrow is a natural builder and instructor who works in the U.S. and Central America. She first started experimenting with clay plasters and later fell in love with lime as well. You can learn more about her work in the U.S. at earthenendeavors.com. She also founded a not-for-profit organization in Central America, working with women and youth in rural communities to improve housing and create community buildings. You can learn more about this exciting project at puebloproject.org. Most recently, she has worked with Vermont Natural Homes in southern Vermont.

Tom and Satomi Lander head LanderLand, their natural and conventional design, consult, and building business. They have been involved with the natural building movement for over 25 years. They also teach and educate through hands-on workshops and specialize in natural earth and lime plastering and restoration. landerland.com

Tom and Satomi also have a mail order business importing and selling high-quality Japanese-made earth and lime plastering tools. They will soon be celebrating their 15th year. They have the largest selection of Japanese trowels in stock in North America, and also offer custom-made Japanese trowels and hawks of their own designs. JapaneseTrowels.com

Deirdre McGahern owns and runs Straworks, Inc., a straw bale building company based in Peterborough, ON, that designs and builds one-of-a-kind, super-insulated homes that are made of natural, local, and nontoxic materials. Deirdre started Straworks in 2004, and 13 years later she still loves her job and employing the people who work for Straworks. She takes pride in their work and has fun doing it, because, when built properly, straw works!

straworks.ca

New Frameworks was founded in 2006 to design, construct, and renovate buildings in a way that prioritizes mitigating climate change and promoting human and ecological health in the built environment. Their buildings connect people to the natural world and to local community through the use of regional forest, geologic, and agricultural products that not only have a low carbon footprint in their production but perform well in energy efficient building designs. New Frameworks is a worker-owned cooperative that focuses on the health, empowerment, and fulfillment of our owners as well as our clients and our community. Get more information at: newframeworks.com

Ian Redfern trained as a mechanical engineer, then spent the next 20 years restoring early Inner Auckland Villas, before taking time out and living in New Mexico, where he fell in love with the ambience and simplicity of adobe homes.

Following his return to New Zealand, he built several adobe homes, then established a boutique Architectural Design Practice devoted to *eco-healthy, sustainable* home designs using earthen, straw bale, and durable timber materials and simple building techniques; he also mentored many owner-builders as they built their homes and ran hands-on workshops.

Ian has retired to Australia with his sculptress wife, where he offers a consulting and advisory service to architects. Contact Ian at ian@studioredfern.com

An innovative builder, plasterer, and artist, ***Tom Rijven's*** colorful personality is a match for his colorful plasters. Nomadic in spirit, Tom is an avid educator of natural building, and is extremely passionate about what he does. Author of *Between Earth and Straw,* Tom continues to expand and develop his own ideas for the best ways of building, and he shares them willingly with others. Whether in Spain, France, or Holland, Tom is sure to have his hands deep in a building project. habitatvegetal.com

The founder of The Lime Plaster Company, heritage plasterer ***Benjamin Scott*** was born and raised in southern England. He began his heritage plastering career in beautiful Chagford,

Devon, through an on-site building apprenticeship, learning everything from traditional brick, block, and stone work, to laying floors and constructing roofs. He quickly turned his attention to the heritage plastering aspect of the work, where he began to specialize in the use of pure lime and hydraulic lime heritage plasters, stuccos, and mortars.

Now, with over 15 years of heritage plaster and heritage mortar construction experience behind him — using many different types of traditional materials, including both hydraulic and non-hydraulic limes, gypsum, and clays — Ben has brought his skills to Canada, where he has focused on expanding his successful heritage plastering business through application and education. For more information, go to: naturallimeplaster.ca

Athena Swentzell Steen grew up in Santa Fe and at the Santa Clara Pueblo, where she began building with natural materials at a very early age. Co-founder of the Canelo project, and co-author of *The Beauty of Straw Bale Homes, Small Strawbale,* and *Built by Hand,* Athena continues to inspire and delight natural builders the world over through her artistic plastering techniques and informative, hands-on workshops, and colorful books. caneloproject.com

Errol Towers is a second-generation drywall and painting contractor living in the White Mountains of central New Hampshire. On the job site since age 10, Errol has spent the majority of his lifetime honing his craft. Wanting to find healthier and more environmentally friendly ways to expand his craft, he has begun to move toward working with both clay- and lime-based plasters. Already recognized as an authority on natural plasters and natural building in his home state, Errol regularly is consulted on projects all across the country. Some of Errol's favorite projects include straw bale homes and numerous Shikkui Japanese lime plaster projects. Marmorino, Stuc Pierre, and ornamental gypsum moldings round out the list.

facebook.com/errol.towers

Catherine Wanek is a co-founder of Builders Without Borders. A passionate advocate for natural building for over two decades, she is the author and photographer of *The Hybrid House* and *The New Straw Bale Home,* and was co-editor/photographer of *The Art of Natural Building.* She also produced the videos *The Strawbale Solution,* the *Building With Straw* video series and *Urban Permaculture.* Catherine is the owner of the Black Range Lodge, a historic bed-and-breakfast inn located in the mountains of southwest New Mexico, which is a center for ecological building and healthy living. BlackRangeLodge.com

Index

Page numbers in *italics* indicate tables.

About the Authors

MICHAEL ***HENRY*** has researched plasters and plastered his way across Ontario for the past decade, plastering for Camel's Back Construction and Straworks. His attention to detail and mad-scientist plaster experiments have made him a noted expert in the field and a sought-after workshop leader on plasters at the Endeavour Centre. Michael is co-author of *Ontario's Old-Growth Forests*, and he lives in Peterborough, Ontario, with his wife and two children. He shares his plastering knowledge at thesustainablehome.net.

Tina Therrien started plastering in 1997 as part of Camel's Back Construction, the first straw bale building company in Ontario. One of the founding members of the Ontario Natural Building Coalition, Tina has made numerous contributions in the natural building world and has plastered in France and Haiti. Passionate about food, gardening, and chickens, Tina lives in a modest timber frame home with her spouse, daughter, their flock of chickens, and their slowly expanding gardens. She is co-author of *More Straw Bale Building*, and she operates Shelter By Hand, a timber framing company, with her spouse. She lives in Low, Quebec.

A Note About the Publisher

NEW SOCIETY PUBLISHERS is an activist, solutions-oriented publisher focused on publishing books for a world of change. Our books offer tips, tools, and insights from leading experts in sustainable building, homesteading, climate change, environment, conscientious commerce, renewable energy, and more — positive solutions for troubled times.

We're proud to hold to the highest environmental and social standards of any publisher in North America. This is why some of our books might cost a little more. We think it's worth it!

- We print all our books in North America, never overseas
- All our books are printed on **100% post-consumer recycled paper**, processed chlorine free, with low-VOC vegetable-based inks (since 2002)
- Our corporate structure is an innovative employee shareholder agreement, so we're one-third employee-owned (since 2015)
- We're carbon-neutral (since 2006)
- We're certified as a B Corporation (since 2016)

At New Society Publishers, we care deeply about *what* we publish — but also about how we do business.

Download our catalogue at https://newsociety.com/Our-Catalog or for a printed copy please email info@newsocietypub.com or call 1-800-567-6772 ext 111

New Society Publishers

ENVIRONMENTAL BENEFITS STATEMENT

For every 5,000 books printed, New Society saves the following resources:[1]

44	Trees
4,007	Pounds of Solid Waste
4,408	Gallons of Water
5,750	Kilowatt Hours of Electricity
7,284	Pounds of Greenhouse Gases
31	Pounds of HAPs, VOCs, and AOX Combined
11	Cubic Yards of Landfill Space

[1]Environmental benefits are calculated based on research done by the Environmental Defense Fund and other members of the Paper Task Force who study the environmental impacts of the paper industry.